INFECTIOUS DISEASES AND MICROBIOLOGY

RECENT TRENDS IN UNDERSTANDING AND TACKLING GRAM-NEGATIVE INFECTIONS

INFECTIOUS DISEASES AND MICROBIOLOGY

Additional books and e-books in this series can be found on Nova's website under the Series tab.

INFECTIOUS DISEASES AND MICROBIOLOGY

RECENT TRENDS IN UNDERSTANDING AND TACKLING GRAM-NEGATIVE INFECTIONS

JAYAPRADHA RAMAKRISHNAN
AND
THIAGARAJAN RAMAN
EDITORS

Copyright © 2020 by Nova Science Publishers, Inc.

All rights reserved. No part of this book may be reproduced, stored in a retrieval system or transmitted in any form or by any means: electronic, electrostatic, magnetic, tape, mechanical photocopying, recording or otherwise without the written permission of the Publisher.

We have partnered with Copyright Clearance Center to make it easy for you to obtain permissions to reuse content from this publication. Simply navigate to this publication's page on Nova's website and locate the "Get Permission" button below the title description. This button is linked directly to the title's permission page on copyright.com. Alternatively, you can visit copyright.com and search by title, ISBN, or ISSN.

For further questions about using the service on copyright.com, please contact:
Copyright Clearance Center
Phone: +1-(978) 750-8400   Fax: +1-(978) 750-4470   E-mail: info@copyright.com.

NOTICE TO THE READER

The Publisher has taken reasonable care in the preparation of this book, but makes no expressed or implied warranty of any kind and assumes no responsibility for any errors or omissions. No liability is assumed for incidental or consequential damages in connection with or arising out of information contained in this book. The Publisher shall not be liable for any special, consequential, or exemplary damages resulting, in whole or in part, from the readers' use of, or reliance upon, this material. Any parts of this book based on government reports are so indicated and copyright is claimed for those parts to the extent applicable to compilations of such works.

Independent verification should be sought for any data, advice or recommendations contained in this book. In addition, no responsibility is assumed by the Publisher for any injury and/or damage to persons or property arising from any methods, products, instructions, ideas or otherwise contained in this publication.

This publication is designed to provide accurate and authoritative information with regard to the subject matter covered herein. It is sold with the clear understanding that the Publisher is not engaged in rendering legal or any other professional services. If legal or any other expert assistance is required, the services of a competent person should be sought. FROM A DECLARATION OF PARTICIPANTS JOINTLY ADOPTED BY A COMMITTEE OF THE AMERICAN BAR ASSOCIATION AND A COMMITTEE OF PUBLISHERS.

Additional color graphics may be available in the e-book version of this book.

Library of Congress Cataloging-in-Publication Data

ISBN: 978-1-53618-503-4
Names: Ramakrishnan, Jayapradha, editor. | Raman, Thiagarajan, editor.
Title: Recent Trends in Understanding and Tackling Gram- Negative Infections/
 [edited by] Jayapradha Ramakrishnan and Thiagarajan Raman,
Jayapradha Ramakrishnan, Senior Assistant Professor,
Centre for Research in Infectious Diseases (CRID),
SASTRA Deemed to be University, Thanjavur, Tamil Nadu, India.
Thiagarajan Raman, Assistant Professor
Department of Advanced Zoology and Biotechnology
Ramakrishna Mission Vivekananda College, Mylapore Chennai, India
Description: New York : Nova Science Publishers, [2020] | Series:
 Infectious diseases and microbiology | Includes bibliographical
 references and index. |
Identifiers: LCCN 2020038448 (print) | LCCN 2020038449 (ebook) | ISBN
 9781536185034 (paperback) | ISBN 9781536185744 (adobe pdf)
Subjects: LCSH: Klebsiella. | Antibiotics--Development. |
 Antibiotics--Physiological effect.
Classification: LCC QR82.E6 I46 2020 (print) | LCC QR82.E6 (ebook) | DDC
 615.7/922--dc23
LC record available at https://lccn.loc.gov/2020038448
LC ebook record available at https://lccn.loc.gov/2020038449

Published by Nova Science Publishers, Inc. † New York

Contents

Preface   vii

Chapter 1   Natural Compounds: A Rich and Renewable Source of Immunomodulators and Antimicrobial Compounds   1
Sudarshan Singh Rathore, Jaya Gangwar, Jayapradha Ramakrishnan and Thiagarajan Raman

Chapter 2   Recent Advances in the Understanding and Management of *Klebsiella pneumoniae* Infections   49
Muhsin Jamal, Sayed Muhammad Ata Ullah Shah Bukhari, Sana Raza, Liloma Shah, Redaina, Kanwal Mazhar and Saadia Andleeb

Chapter 3   Antibiotic Resistance in *Klebsiella pneumoniae*   91
Suganthi Rasangam, Aris Chandran Abdullah and V. Gopalakrishnan

Chapter 4	Macrophage Immunomodulation by Biomolecules for Eradicating *Pseudomonas aeruginosa* R. Sangeetha, D. Parimalanandhini, K. Mahalakshmi, M. Livya Catherene, M. Beulaja, R. Thiagarajan, S. Janarthanan, M. Arumugam and R. Manikandan	105
Chapter 5	Survival Strategies of *Pseudomonas aeruginosa* in Ocular Infections Vidyarani Mohankumar, Kathirvel Kandasamy, Prajna Lalitha and Bharanidharan Devarajan	125
Chapter 6	Mechanism of Quorum Sensing and Inhibition of Quorum Sensing Mediated Virulence by Putative Natural Compounds against Gram-Negative Bacterial Pathogens A. Annapoorani, R. Manikandan, A. Veera Ravi and S. Janarthanan	143
About the Editors		183
Index		185

PREFACE

Antibiotics have vastly changed the way we fight diseases. History is replete with examples of infectious diseases that have killed millions of people worldwide in the past and continues to do so in the present. However, there is one major difference. In the past there were novel classes of antibiotics that were being discovered at regular intervals, and in the present we are running out of options. Moreover, whatever antibiotics we have, microbes have developed resistance, which could primarily be attributed to the injudicious usage of antibiotics, not only for humans but also for veterinary purposes. A previous report from World Health Organization (WHO, 2017) suggests that antimicrobial resistance is a serious hazard and antibiotics under clinical development may not be sufficient (or efficient enough) to treat these emerging resistant pathogens. That report had identified 51 antibiotics and 11 biologicals. Of those, only 12 were found to be active against WHO classified critical priority pathogens and only two of them were found to be active against more than one specific pathogen. Starting with the sulphonamides in the 1930's to lipopeptides in the early 2000's, research and development on novel antibiotics is usually a slow and painful process and currently it will suffice to say that it is on the decline. As per the Infectious Diseases Society of America (IDSA), we are facing what is known as an "antibiotic paradox" that is pushing against the development of novel antibiotics. Interestingly, though there are studies that keep reporting on antibiotics, most of these are on combinatorial use of antibiotics. What could be their effectiveness and more importantly what could be their biotoxicity in the long run, remains unknown. Microbial antibiotic resistance is not a new phenomenon and there are numerous studies that have demonstrated the various mechanisms underlying it. An interesting study by Indian Council of Medical Research (ICMR), has shown that commensals in the gastrointestinal tract of humans might be responsible for the increasing ineffectiveness of antibiotics, at least in Indians. This makes it clear that microbes will keep on producing resistance to antibiotics at a faster rate compared to our ability to develop them. Under these circumstances, natural compounds, primarily plant-based, have become valuable tools and could be our answer to not only effective antimicrobial principles, but also to antimicrobial resistance. These compounds are abundant in nature and there is already a very rich literature on their usage and efficacy based on the various traditional systems of medicine. These natural compounds have been shown to be effective against both Gram positive and Gram-negative pathogens and interestingly microbes have a limited chance (as far as studies show) of developing resistance towards them. This is primarily attributed to the fact that most of these compounds are used as polyherbal formulations. Another advantage in studying these natural compounds is that there is a better probability of hitting upon that 'jackpot' molecule or molecules for antimicrobial applications. Being relatively safe and inexpensive makes them very attractive areas for clinical research. This book, in line with others in the field, is a small attempt to highlight the developments related to the antimicrobial compounds from natural sources and their mechanisms of action, particularly against ESKAPE pathogens. This book, we believe, will serve as a small but important piece of source material for students and researchers interested in this particular area of research. The chapters are divided to showcase the relevance and importance of natural compounds as novel antimicrobials, inhibitors of antimicrobial resistance and immunomodulators and we hope that the topics will kindle the interest of young researchers in these lines. In the end, we are grateful and whole heartedly acknowledge the authors for their valuable contribution and reviewers for their valuable suggestions and critical review of the manuscripts.

Jayapradha Ramakrishnan
Thiagarajan Raman

In: Recent Trends in Understanding …
Editors: J. Ramakrishnan et al.
ISBN: 978-1-53618-503-4
© 2020 Nova Science Publishers, Inc.

Chapter 1

NATURAL COMPOUNDS: A RICH AND RENEWABLE SOURCE OF IMMUNOMODULATORS AND ANTIMICROBIAL COMPOUNDS

Sudarshan Singh Rathore[1], Jaya Gangwar[1], Jayapradha Ramakrishnan[1,] and Thiagarajan Raman[2,†]*

[1]Actinomycetes Bioprospecting Lab, Centre for Research, in Infectious Diseases (CRID), SASTRA Deemed to be University, Tirumalaisamudram, Thanjavur, Tamil Nadu, India
[2]Department of Advanced Zoology and Biotechnology Ramakrishna Mission Vivekananda College, Mylapore Chennai, India

ABSTRACT

The immune system of the human is one of the most complex biological systems to protect our body from various diseases and their

---

[*] Corresponding Author's Email: antibioticbiology@gmail.com.
[†] Corresponding Author's Email: thiagi200@yahoo.co.in.

damage. Immunity can be divide into two types based on its response, active and passive immunity. Active immunity is body response against disease but if external bioactive compounds involved in modulating the immune system define passive immunity.

Immune system dysfunction is responsible for various life-threatening diseases like asthma, allergy, cancer, and antibiotic-resistant infectious pathogen, etc. but the use of bioactive immunomodulators help to manage diseases by modulating the immune response. Bioactive compounds with immunomodulatory potentials are the next generation drugs capable of modulating the immune system against various diseases. Many of the marine and terrestrial originated plants, animals, and microorganisms derived bioactive biomaterial such as protein; peptides, polysaccharides, etc. have shown to modulate the immune system by different mechanisms, and each trial these compounds significantly regulate the immune system. These bioactive biomaterials have yet to acknowledge the same as novel pharmaceuticals. Identified immunomodulatory bioactive compounds have extracted from different biological sources were used as traditional medicines in various countries in the past. These traditional medicines may act as base literature for future research to cure various complex and untreatable diseases. Also, it will promote physical health, mental health, improved immune system, and enhance its response for a longer period with lesser side effects compare to synthetic immunomodulatory compounds. This chapter is highlighting research done on bioactive compounds extracted from various terrestrial and marine sources with immunomodulatory potentials. Also, the chapter discussed the biological relevance and mechanism involved in modulating the immune system. This information may encourage researchers to explore natural bioactive compounds with immunomodulatory potentials.

Keywords: immunomodulation, immunity, bioactive compound, biomaterial, diseases

1. IMMUNOMODULATORS

Immunomodulators immunity plays a major role to protect the host from various diseases and its damage. Host susceptibility defines the disease intensity and damages to the host mechanism. Disease in host depends on various factors such as due to infectious microorganisms, a mutation in host cell cause cancer, etc. Host immune system interaction and response against disease only prevent and reduce the damage in individuals. Host immunity and disease relationship are demonstrated in the immunocompromised host, by disease caused by the foreign or mutated host cell and the successful reconstitution of the immune response to cure of disease and caused damages to host (L. Anne Pirofski and Casadevall 2006). The immunomodulation approach is useful to comprehend the pathogenesis and host immune response. In contrast to microbial virulent factors and pathogenesis are considering establishing a correlation between immune response and damage response to cure disease (L. A. Pirofski and Casadevall 2018). The misuse or overuse of antimicrobial agents in daily life has lemmatized the new antimicrobial drug research. The best approach is to develop antimicrobial therapy that regulates immune response to reduce antimicrobial resistance and host damage caused by disease and drugs (Spellberg et al. 2004).

Immunology is one of the most promising and rapidly developing areas of biomedical research for the prevention and treatment of a wide range of infectious and non-infectious diseases. The defense system to protect from foreign agents is highly specific and remarkable in vertebrates. Immune response has various ways to tackle infectious and non-infectious diseases. In one of the way T lymphocytes directly kill the cell, induce an inflammatory response, and activate other immune cells is known as cell-mediated immunity. The other system B lymphocytes regulated by a cell-mediated immune response to synthesize specific immunoglobulins are known as humoral immunity (Ramasundaram Srikumar, Jeya Parthasarathy, and Sheela Devi 2005). It comprises varieties of molecules and cells having the capability to recognize/ neutralize and eliminate toxic/undesirable molecules or cells in the body. Biological and synthetic molecules that can alter the immune response by inducing, expressing, amplify, or inhibit adaptive and innate immune response known as immunomodulators. Immunopharmacology is a developing branch of pharmacology to investigate immunomodulators potential use in clinical medication for the reconstitution of immune deficiency for treatment of AIDS and the suppression of hyperactive or normal immune response for autoimmune disease or organ transplantation (Manu and Kuttan 2009). Immunomodulators are classified into three types based on their effects and responses: Immunoadjuvants; Immunostimu-lants; Immunosuppressants.

Immunoadjuvants are used with vaccines to induce promising and specific enhancement of immune stimulation to begin modulation of the immune response. It will establish a connection between cellular and humoral immune response in the host (Alfons B, Patrick M 2001).

Immunostimulants are enhancing innate and adaptive immune responses to a body against infection. In a healthy individual, they act as an immunotherapeutic agent and expected to serve as promoters or prophylactic agents (Juyal PD, Singla LD).

Immunosuppressants are a group of drugs that are used to treat autoimmune diseases and to protect organ transplant rejections (El-Sheikh 2008).

2. ROLE OF IMMUNOMODULATORS

The immunomodulation approach can be segregated based on pathogen-specific and non-specific. Antibody and vaccines are pathogen-specific immunomodulators whereas probiotics, antimicrobial peptides, antibiotics, and cytokines are non-specific immunomodulators. Currently licensed vaccines are used as a specific immunomodulator to prevent diseases rather than treatment. Similarly, cytokines are in clinical practice as non-specific immunomodulators to treat infectious diseases and natural products use as a supplement as an immunomodulator and a fast recovery of damages due to diseases.

In early 20th century therapy for infectious diseases based on antibody treatment prepared from serum and it was the inaugural of a specific immunomodulator as an antimicrobial agent (A. Casadevall 2006). First-generation of antibody therapy was not successful due to impurities or derived from non-human species. After the arrival of antimicrobial compounds with immunomodulatory potentials which were acted directly on infectious microorganisms were classified as non-specific immunomodulators. During this era antimicrobial compounds mode of action was unknown but the serum therapy validation in animal models was established before administered to humans (A. Casadevall and Scharff

1994). Mechanism of antibody treatment was thought that the antibody opsonizes the pathogen activate the complement system and enhances the phagocytosis or antibody-dependent cell-mediated cytotoxic response, also antibody neutralizes antigens and toxins to reduce the damage in the host (Arturo Casadevall and Pirofski 2004). After some time potential mechanism of action of antibody was documented in which the generation of oxidative species by immune cells directly kills the pathogen and modulates the immune system. For example, pulmonary infection by pneumococcus in the lung of mice treated with serum contains capsular polysaccharide-specific antibody modulates the cell-mediated immune response. This study suggested that serum therapy with a specific antibody can reduce the clinical symptoms and damage caused by pneumococcal pneumonia by downregulating the host inflammatory response. Also, the study suggested that serum therapy was limited to the early stage of the disease, which means therapy was successful only when started within the first three days of symptoms with the capsular polysaccharides- specific antibody (Burns, Abadi, and Pirofski 2005; A. Casadevall and Scharff 1994)(A. Casadevall and Scharff 1994). Antibody therapy was demonstrated for other microorganisms also, which suggested that it was effective against extracellular pathogens only, whereas for intracellular pathogens cell-mediated immunity was effective. Few studies suggested that antibody serum therapy effective against classical intracellular pathogens such as *Cryptococcus neoformans*, *Mycobacterium tuberculosis*, etc. In this case, antibody opsonization helped cell-mediated immunity to recognize the pathogen faster than untreated with a specific antibody (Arturo Casadevall and Pirofski 2006). Non-pathogen specific cytokine based therapies for infectious diseases is a natural mediator to enhance the antimicrobial effect of host immune response with or without antimicrobial agents (L. anne Pirofski and Casadevall 2006).

Curcumin is well known and approved antimicrobial compound that has anti-inflammatory potentials to cure infectious disease by a synergistic approach. The study was carried out on curcumin by Bansal et al. 2013 in *Klebsiella pneumoniae* infected BALB/c mice to protect from lung inflammation (Catanzaro et al. 2018). In this study, curcumin was administered by oral route either alone or in combination with supplements that showed a significant reduction of infiltrating neutrophils in the alveolar wall of lungs and a decrease in the production of TNF-α, NO and MPO activity (Catanzaro et al. 2018).

3. MECHANISM OF ACTION OF THE IMMUNOMODULATORS

The bioactive compound from the natural source has various properties along with immunomodulation such as antioxidant, anti-cancer, anti-diabetic, anti-aging, effective for parkinson's disease, rheumatoid arthritis, and autoimmune disease. These compounds modulate the immune response through suppression, stimulation, or adjuvant (Poudel, Pradeep, and Yadav 2019). Mechanism of immunomodulatory compounds mainly stimulates phagocytosis, activates macrophage, non-specific cell-mediated immune function enhancement, increase antigen-specific antibody production, increase circulating immune cell count, and stimulatory cytokines levels (Y. K. Wang et al. 2010a) (Khan et al. 2012) (Bharani et al. 2010) (Panax ginseng 2009) (Long 2016). To maintain the disease-free state in the normal or unhealthy individual by modulating the immune response by stimulating or suppressing. Natural compounds that activate the host immune system are supportive therapy with conventional chemotherapy which often damages immune response (Harborne 1987). Cytotoxic drugs target proliferative cells also target, highly sensitive bone marrow stem cells. It will result in thrombocytopenia and leucopenia, due to the effect on the bone marrow to regenerate new blood cells. 96 A large number of compounds such as flavonoids, curcumin, tannic acid, tocopherol, polyphenols, ascorbate, carotenoids, etc., have been reported as a potential immunomodulatory compound. Indian traditional medicine is a mixture of herbal compounds with immunomodulatory potential act synergistically to cure disease without any toxicity. This will be important to understand the use of a compound in the appropriate concentration. Pharmacological studies have been performed at *in vitro* and *in vivo* to screen the different types of compounds and their effects (Harborne 1987) (D. Kumar et al. 2012).

4. NATURAL SOURCES FOR IMMUNOMODULATORS

The immune system is responsible for the host health by eliminating infectious agents, aging cells, or tumor cells from the body, but its response can be dependent on various surrounding factors such as pathogenic load, tissue injury, etc (Villani, Sarkizova, and Hacohen 2018). The immune system is having the ability to modulate its response and protect it from many life-threatening diseases (Routy, Mehraj, and Cao 2016) (Naidoo, Page, and Wolchok 2014) (Havla, Kümpfel, and Hohlfeld 2015). Mononuclear leukocytes such as macrophages are the primary defender of the host immune system, which can recognize and eliminate infectious microorganisms and mutated cancer cells from the host via phagocytosis. Macrophages regulate the immune system by produce cytokines such as interleukins, interferons, and tumor necrosis factors to protect from disease and damage. Currently, many immunomodulatory drugs are used clinically to control immune response against and diseases. Cyclosporine A, imiquimod, cyclophosphamide, tilorone, penicillamine, thiocarbamate, pidotimod, niridazole, levamisole, and prostaglandin are few immunomodulatory drugs (Gruppen et al. 2018) (Ulrich et al. 2007)(Trabattoni et al. 2017)(Gruppen et al. 2018)(Ekins et al. 2018) (Ahlmann and Hempel 2016) (Cornejo-García et al. 2016) (Flores et al. 2019) (Jantan, Ahmad, and Bukhari 2015) (Jantan, Ahmad, and Bukhari 2015) (Ozkan et al. 2015) (Gong, Klingenberg, and Gluud 2004)Compare to naturally derived immunomodulatory drugs synthesized drugs have toxicity, side effects, and high cost based on their dose and usage (Y. K. Wang et al. 2010a). Several bioactive compounds from the microbial, plant and animal sources have immunomodulatory potentials along with other properties like antioxidant, anti-inflammatory, antimicrobial, hepatoprotective, cardiotonic, hypocholesterolemic and other bioactive properties.

4.1. Plant-Derived Immunomodulators

Plants and herbs are providing food, shelter, and bioactive compounds to cure various diseases. The three-quarters of the world population depends on traditional remedies to cure various diseases. Herbal or traditional natural medicine has used in various countries such as Ayurvedic (India), Chinese, Kampo (Japan), Greco-Arab, Unani-Tibb (South Asia), Egyptian, etc. Natural traditional therapy is currently being in practice through extensive research activity on various plant species and properties (D. Kumar et al. 2012). Plant extracts are a combination of its ability to inhibit pathogenic microorganisms growth and support immune function in various parts of the body. These extracts can improve cell-mediated and humoral immunity against bacteria, fungi, viruses, and parasites (Vinothapooshan and Sundar 2011). The following are the few examples of immunomodulatory compounds.

4.1.1. Glycosides

Glycosides are a combination of one or more sugar components get by enzymatic or acid hydrolysis of an organic compound from animal or plant sources. They are acetals or sugar ethers, formed by sugar and non-sugar entities hydroxyl group interactions, with the loss of water molecule. Many glycosides have been reported as an immunomodulatory potential. Sesquiterpene glycosides like Dendronobilosides A and B, and Dendroside A are used in Chinese traditional medicine have been isolated from *Dendrobium nobile* plant species. Dendronobilosides A and Dendroside A were reported in *in vitro* studies to stimulate the proliferation of murine T and B lymphocytes whereas Dendronobilosides B showed inhibitory activity. Iridoid glycosides isolated from *Picrorhiza scrophulariiflora* reported as inhibitory activity towards the classical pathway of complement and antioxidant (Smit HF 2000). Anthraquinone glycosides isolated from *Andrographis paniculata* reported as Hepatoprotective, antispasmodic, blood purifier, febrifuge (*Indian Medicinal Plants* 2007)(Shrivastava, Varma, and Padh 2011).

Coumarins are derivatives of glycosides are also have immunomodulatory potentials. Coumarins isolated from *Achillea millefolium* and *Apium graveolens* have anti-inflammatory potentials (Sharififar et al. 2009) (Morris et al. 2007). Esculetin (6,7-dihydroxycoumarin) isolated from many plants such as *Euphorbia lathyris*,

Citrus limonia, Artemisia capillaries have an anti-inflammatory, immunostimulatory effect on lymphocyte proliferation, induction of cytokine such as TNF-α, INF-γ) (Tanaka et al. 1999) (Noori et al. 2004) (SM et al. 2009).

4.1.2. Flavonoids

Flavonoids extracted from *Achillea millefolium, Alternanthera tenella* have reported as immunomodulatory potential as an anti-inflammatory (Sharififar et al. 2009) (McFadden et al. 2003). Other flavonoids extracted from *Terminalia arjuna, Bauhinia variegata, Abutilon indicum* such as apigenin, oligomeric proanthocyanidins, isoflavonoids, flavones, and anthocyanidins are reported as an antioxidant, antimicrobial, cardiotonic, diuretic, prescribed for hypertension (Silva et al. 2016) (Ghaisas, Shaikh, and Deshpande 2009) (Sharififar et al. 2009).

4.1.3. Alkaloids

Alkaloids are natural or synthetic heterocyclic organic compounds reported as immunomodulatory compounds with antimicrobial activity. Alkaloids isolated from Murraya koenigii, Actinidia macrosperma, Achillea millefolium, and Cissampelos pareira. It has anti-inflammatory, antimicrobial, Antipyretic, analgesic, antilithic potentials were reported (Sharififar et al. 2009) (Morris et al. 2007)(Pradhan, Panda, and Tripathy 2009) (Alamgir and Uddin 2010).

4.1.4. Polysaccharides

Polysaccharides have numerous therapeutic potentials and modulating innate immune response mainly macrophage activation. Polysaccharides can be extracted from microbial or plant sources, but both the polysaccharides have common receptors and responses in macrophages. Extracted polysaccharides have the potential to activate macrophages and enhance the proliferation of T lymphocytes. Bioactive high molecular weight substance extracted from *Salicornia herbacea* used to treat various diseases including cancer. Also, it can activate and differentiate monocytic cells to macrophages (Alamgir and Uddin 2010)(Ríos 2010)

4.2. Marine-Derived Immunomodulators

After extensive research on terrestrial plants and microorganisms for the bioactive compound, the last few decades researchers focus to isolate compounds from marine organisms. More than 16,000 marine species were widely studies isolated from oceans. Marine organisms derived proteins, glycoproteins, peptides, lipids, and polysaccharides show immunomodulatory potentials and other bioactive properties such as antimicrobial, anticancer, etc (Kiewiet, Faas, and de Vos 2018)(Okolie et al. 2017) (Miccadei et al. 2016).Marine-derived immunomodulatory compounds and their effect on disease control are as follow:

4.2.1. Immunomodulatory Amino Acids and Proteins

Marine natural sources were used to extract various marine biomaterials such as proteins, enzymes, fatty acids, oligosaccharides, pigments, biopolymers, etc. A major portion of marine biomaterials contains a verity of proteins (10 to 47%) with bioactive potentials and functions. Proteins in hemocytes and hemolymphs are preserving various immune components and play a major role in innate immunity such as glycoprotein, metalloproteins, antimicrobial peptides, coagulation factors, protease inhibitors, and amino sulfonic acid. The following are a few of the immunomodulatory proteins found in marine biomaterials (Raoult 1993).

4.2.1.1. Hemocyanins

Hemocyanins are metalloproteins that found in gastropods such as mollusks to transport oxygen by binding with metals like copper. It shows immunostimulatory potential against few cancer types with minimal toxicity in murine colon carcinoma model (Dolashka-Angelova et al. 2008) (Del Campo et al. 2011). Hemocyanin extracted from *Concholepas concholepas* stimulates innate immunity to suppress the growth of superficial bladder cancer. Keyhole limpet hemocyanin is isolated from *Megathura crenulata* marine mollusk circulating glycoprotein, which enhances host immune response by interacting with T lymphocytes, monocytes, and macrophages (Lammers et al. 2012) (Román et al. 2019). Similarly, hemocyanin isolated from *Fissurella latimarginata* gastropod reported higher immunostimulatory potentials to compare to other hemocyanins (Arancibia et al. 2014).

4.2.1.2. Lectins

Lectins are glycoprotein that controls innate immune response by pathogen recognition and phagocytosis, also induce an adaptive immune response by activating B and T lymphocytes (Sharon 2007). Lectins are isolated from numerous plants and animals and classified into three types: C-type (Ca^{2+}-dependent), R-type, and metal independent galectins. Tachylectin extracted from *Carcinoscorpius rotundicauda* horseshoe crab help in pathogen recognition and enhances phagocytosis or lectin pathway in complement system (Ma et al. 2004). Lectins isolated from *Mytilus trossulus* mussel stimulate the production of pro-inflammatory cytokines such as TNF-α and IFN-γ but suppress the production of anti-inflammatory cytokines such as IL-10 (Chikalovets et al. 2013). Lectins isolated from hemolymph play a major role in B lymphocyte activation to produce antibodies for pathogen recognition and antimicrobial activity (Fredrick and Ravichandran 2012).

4.2.1.3. Taurine

Taurine (2-amino ethane sulfonic acid) is an amino acid is found in animal tissue, marine clam, and other marine organisms. Taurine is cytoprotective and has immunomodulatory potentials. It enhances the efficacy of immune cells such as neutrophils, monocytes, and lymphocytes (Marcinkiewicz and Kontny 2014). Taurine reported its role in innate immunity by phagocytosis mediated inflammatory response at the site of infection. Taurine accumulates in phagocytes and activates macrophages to produce cytokines, oxidants, and peroxidants to activate immune cells and kill the pathogen by the inflammatory response (Schuller-Levis and Park 2004). cytoprotective property of taurine protects host immune cells from oxidants dependent on inflammatory damage. Taurine activates peroxisome proliferator-activated receptor-γ (PPAR-γ) in the liver to regulate glucose metabolism in a diabetic individual which protects from retinal and neuronal damage (M. K. Song et al. 2011).

4.2.2. Immunomodulatory and Antimicrobial Peptides

Immunomodulatory potentials of marine peptides are studied to enhance the immune response in the host. These peptides are known to activate natural killer cells, lymphocytes proliferation, and cytokines regulation. The mode of action on host systems like digestive, cardiovascular, nervous, and the immune system is not well understood (Murray and FitzGerald 2007) (B. P. Singh, Vij, and Hati 2014) (Y. Wang et al. 2018).Hence, it is a real need to explore the efficacy of peptides and their interaction with host systems to develop an understanding of immunomodulatory peptides. Marine-derived antimicrobial peptide molecules have the potential to enhance innate immunity with target-specific antimicrobial activity as a drug candidate (Xu et al. 2010) (Kang, Seo, and Park 2015). Marine-derived AMPs compound with its efficacy and immunomodulatory potentials are as follow:

4.2.2.1. Callinectin

Callinectin is an antimicrobial peptide isolated from *Callinectes sapidus* (hemocytes of the blue crab) and *Mytilus galloprovincialis* is Mediterranean mussel. It shows antibacterial activity against gram-negative bacteria and composed of 32 amino acids rich in proline, arginine, and four cysteine residues. N-formylkynurenine, hydroxy-N-formylkynurenine group, and hydroxyl tryptophan are callinectin isoforms have changes in tryptophan residues. Callinectin binds to blue crab hemocytes anti-callinectin-like peptide antibodies and shows antimicrobial and immunomodulatory potentials (Noga et al. 2011) (Li et al. 2010).

4.2.2.2. Clavanin A and Clavanin-MO

Clavanin A shows broad antimicrobial activity was isolated from hemocytes of *Styela clava* is a marine tunicate. It has shown activity against fungi, MDR Gram-positive, and Gram-negative bacteria in both *in vitro* and *in vivo* studies (Hee Lee, Cho, and Lehrer 1997). Clavanin-MO is a synthetic derivative of Clavanin A by substituting hydrophilic amino acids at Nterminus to enhance its cell interaction and cell membrane penetration ability compare to Clavanin A. Clavanin-MO have antibacterial, antibiofilm activity against Gram-positive and Gram-negative bacteria also shows anti-inflammatory response caused during infection and host damage. As an immunomodulatory response, these peptides increase anti-inflammatory cytokine (IL-10) levels and decrease pro-inflammatory cytokines (IL-12 and TNF-α) which control the inflammatory damage caused by an immune response against a pathogen (Silva et al. 2016).

4.2.2.3. Crustin

Crustins is antimicrobial peptide rich in cysteine with a typical whey acidic protein (WAP) domain isolated from crustaceans such as *C. maenas*, *Penaeus monodon*, *Scylla paramamosain*, and *Pacifastacus leniusculus* (MW 7–14 kDa). It plays an important role in innate immunity against marine Gram-positive bacteria *Corynebacterium glutamicum*. Crustaceans hemocytes release crustins by exocytosis (Smith et al. 2008) (Donpudsa et al. 2010) (Suleiman, Smith, and Dyrynda 2017)(Imjongjirak et al. 2009).

4.2.2.4. Defensin

Defensins are antimicrobial peptides rich in cysteine found in marine plants, animals and insects act as a host defense peptide. Based on sequence and cycteine residues defensins are classified into two types α and β defensins. Defensins were isolated from *Crassostrea virginica* acidifies gill extract, oysters (*C. gigas* and *C. virginica*), and mussels (*Mytilus edulis* and *M. galloprovincialis*) (Yang et al. 2004a, 2004b) (Zhu and Gao 2013a)(Hubert 1996a). Defensins disrupt the microbial pathogen cell membrane and modulating innate and adaptive immune response (Seo et al. 2005).

4.2.2.5. Myticin

Myticin is antimicrobial peptides rich in cysteine found in mussel *Mediterranean galloprovincialis* in three isoforms myticin A, B, and C (MW 4.4 – 4.6 kDa). Myticin isolated from plasma and hemocytes of *M. galloprovincialis* and have activity against Gram-positive bacteria, Gram-negative bacteria (*E. coli*), fungus (*Fusarium oxysporum*) and also have a role in innate immunity response (Mitta et al. 1999) (Balseiro et al. 2011)(Domeneghetti et al. 2015).

4.2.2.6. Mytilin

Mytilin antimicrobial peptides are rich in cysteine found in marine mollusks. Mytilin has five isoforms mytilin A, B, C, D and G1 were extracted from *Mytilus galloprovincialis* and *M. edulis*. Mytilin shows antimicrobial activity and enhances phagocytosis to protect the host from the spread of infection (Charlet et al. 1996a) (Mitta et al. 2000).

4.2.2.7. Mytimycin

Mytimycin shows antifungal activity consists of 12 cysteines, 6 disulfides connecting bridge, and EF-hand domain (Ca^{2+} binding motif) at C-terminal extension. It mainly expressed in circulating hemocytes and shows antimicrobial activity (Hubert 1996b) (Charlet et al.1996b)

4.2.3. Protein Hydrolysates

Protein derived protein hydrolysates have a wide range of bioactive potentials such as antimicrobial, anticancer, antioxidant, anti-hypertension, and immunomodulatory (Ruiz-Ruiz, I. Mancera-Andrade, and M. N. Iqbal 2016)(Kiewiet, Faas, and de Vos 2018)(Chalamaiah, Yu, and Wu 2018) (Yang et al. 2004b). Peptide bond cleavage in protein during protein hydrolysis results in the formation of different size bioactive peptides. Various proteolytic enzymes such as trypsin, pancreatin, thermolysin, pepsin, KojizymeTM, α-chymotrypsin, Protamex, Neutrase, Alcalase, and Flavourzyme are used to produce immunomodulatory protein hydrolysates. Antimicrobial peptides isolated from marine protein hydrolysates are important and reported in the first line of host defense against pathogenic microorganisms (Yang et al. 2004b) (Zhu and Gao 2013b)(Tang et al. 2015)(R. Song et al. 2012)(Beaulieu et al. 2013) (Balakrishnan et al. 2011). Marine-derived protein hydrolysates with antimicrobial and immunomodulatory potentials are as follow:

4.2.3.1. Chlorella

Chlorella protein hydrolysate prepared by hydrolysis of an ethanol extracted cell biomass of green microalga *Chlorella vulgaris* with pancreatin. Oral administration of chlorella protein hydrolysate in undernourished BALB/c mice supports innate and specific immune response to recover leukocyte count in peripheral blood and bone marrow cellularity. Also, it enhances the production of T cell-dependent antibody reactions, increases lymphocyte pool, and restoration of delayed-type hypersensitivity reaction. Starved mice treated with chlorella protein hydrolysate showed an increased number of activated peritoneal macrophages compare to untreated mice (Morris et al. 2007).

4.2.3.2. Ecklonia

Ecklonia protein hydrolysate prepared by hydrolysis of brown seaweed Alariaceae *Ecklonia cava* cell biomass with Kojizyme. Ecklonia protein hydrolysate shows activation or suppression of the immune cell of murine splenocytes. In ICR mice it enhances splenocyte proliferation and increases the number of monocytes, granulocytes, lymphocytes, and splenocytes. Ecklonia protein hydrolysate increases the number of B cells, CD4+, CD8+ T lymphocytes, and downregulate Th1 cytokines (TNF-α, IFN-γ) and upregulate Th2 cytokines (IL-4, IL-10), Hence Ecklonia protein hydrolysate has an anti-inflammatory immunomodulatory effect (Ahn et al. 2008).

4.2.3.3. Porphyra

Porphyra protein hydrolysate prepared by hydrolysis of algae *Porphyra columbina* with a combination of trypsin and Alcalase. Porphyra protein hydrolysate suppresses the immune response by enhancing the production of anti-inflammatory cytokine (IL-10) and decrease the production of pro-inflammatory cytokines (TNF-α, IFN-γ) (Cian, Martínez-Augustin, and Drago 2012).

4.2.3.4. Porphyra Columbina

Porphyra columbina protein hydrolysate has prepared by hydrolysis of *Porphyra columbina* with flavourzyme and fungal protease concentrate. Porphyra columbina protein hydrolysate has alanine, aspartic acid, and glutamic acid which shows the immunomodulatory effect on T lymphocytes, macrophages, and splenocytes. Porphyra columbina protein hydrolysate enhances the production of IL-10 regulated by p38 MAPK, JNK, and NF-κB pathways in T cells, and inhibits TNF-α, IL-1, and IL-6 production in macrophages (Cian et al. 2012).

4.2.3.5. Edible Red Algae

Edible red algae protein hydrolysate has prepared by hydrolysis of Porphyra tenera with a combination of proteases (Protamex, Flavourzyme, Alcalase, and Neutrase) and carbohydrates (Viscozyme, Promozyme, Maltogenase, amyloglucosidase, Celluclast, Termamyl, and Dextrozyme). These protein hydrolysates show antioxidant and anti-inflammatory response by inhibiting LPS induced nitric oxide production in macrophages. Due to this, it has reported as an anti-inflammatory drug (Vo et al. 2014).

4.2.3.6. Edible Microalgae Spirulina

Edible microalgae spirulina protein hydrolysate has prepared by hydrolysis of filamentous blue-green algae *Spirulina maxima* with pepsin, trypsin, and α-chymotrypsin. Edible microalgae Spirulina protein hydrolysate two peptides MMLDF (655 Da) and LDAVNR (686 Da) shows mast cell degranulation by inhibiting RBL-2H3 by increasing intracellular Ca^{2+} and decreasing histamine release. LDAVNR peptide inhibition block Ca^{2+} and microtubule-dependent signaling pathway. MMLDF peptide inhibition involves ROS production and phospholipase C activation. Inhibition of both the peptides MMLDF and LDAVNR effect on IL-4 production by decreasing NF-κB translocation (Cai et al. 2013).

4.2.3.7. Oyster Peptide-Based Enteral Nutrition Formula

Peptide-based enteral nutrition formula has prepared by hydrolysis of oyster *Crassostrea hongkongensis* with pepsin, trypsin, and bromelain. Lymphocyte proliferation and NK cell activation were observed in cyclophosphamide-induced immunosuppression mice and malabsorption mice treated with Peptide-based enteral nutrition formulation. Oyster peptide-based enteral nutrition formula showed immunostimulatory potentials in mice (Y. K. Wang et al. 2010b).

Table 1. Immunomodulators from Natural sources

S.no.	Class	Source	Activity	References
Plant derived immunomodulator				
1.	Glycosides	Andrographis paniculata	Hepatoprotective, antispasmodic, blood purifier, febrifuge	Shrivastava, Varma, and Padh 2011.
1.1	Coumarins	Achillea millefolium and Apium graveolens	Anti-inflammatory potentials	Sharififar et al. 2009, Morris et al. 2007.
1.2	Esculetin	Euphorbia lathyris, Citrus limonia, Artemisia capillaries	Anti-inflammatory, immunostimulatory effect on lymphocyte proliferation, induction of cytokine such as TNF-α, INF-γ)	Tanaka et al. 1999, Noori et al. 2004, SM et al. 2009.
2.	Flavonoids	*Achillea millefolium, Alternanthera tenella*	Immunomodulatory potential as an anti-inflammatory	Sharififar et al. 2009, Mcfadden et al. 2003
2.1	Apigenin	Adinandra nitida	Lowering blood pressure, and its antibacterial and antiviral properties	Silva et al. 2016, Ghaisas, Shaikh, and Deshpande 2009, Sharififar et al. 2009.
2.2	Oligomeric proanthocyanidins	Pine bark and grape seeds	Antioxidant capabilities greater than those of vitamin C and E, OPC usage might contribute to the improvement of red blood cell function in type 2 diabetes	Silva et al. 2016, Ghaisas, Shaikh, and Deshpande 2009, Sharififar et al. 2009.
2.3	Isoflavonoids	Pueraria lobata (Willd.) Ohwi root (kudzu, Gegen)	Antimicrobial, cardiotonic, diuretic	Silva et al. 2016, Ghaisas, Shaikh, and Deshpande 2009, Sharififar et al. 2009.
2.4	Flavones	Euphorbia neriifolia leaves	Natural pesticides in plants, providing protection against insects and fungal diseases	Silva et al. 2016, Ghaisas, Shaikh, and Deshpande 2009, Sharififar et al. 2009.
2.5	Anthocyanidins	Vaccinium species	Capacity to lower blood pressure, improve visual acuity, reduce cancer cell proliferation, inhibit tumor formation, prevent diabetes, lower the risk of CVD modulate cognitive and motor function	Silva et al. 2016, Ghaisas, Shaikh, and Deshpande 2009, Sharififar et al. 2009.

Table 1. (Continued)

S.no.	Class	Source	Activity	References
3.	Alkaloids	Murraya koenigii, Actinidia macrosperma, Achillea millefolium, and Cissampelos pareira	Anti-inflammatory, antimicrobial, Antipyretic, analgesic, antilithic potentials	Sharififar et al. 2009, Morris et al. 2007, Pradhan, Panda, and Tripathy 2009, Alamgir and Uddin 2010.
4.	Polysaccharides	Microbial or plant sources	Activate macrophages and enhance the proliferation of T lymphocytes	Alamgir and Uddin 2010, Rios 2010.
Marine derived immunomodulator				
Immunomodulatory amino acids and proteins				
1.	Hemocyanins	Concholepas concholepas, Fissurella latimarginata gastropod	Enhances host immune response by interacting with T lymphocytes, monocytes, and macrophages	Dolashka-Angelova et al. 2008, del Campo et al. 2011, Lammers et al. 2012, Román et al. 2019, Arancibia et al. 2014.
2.	Lectins			
2.1	C-type (Ca^{2+}-dependent)	Carcinoscorpius rotundicauda	Pathogen recognition and enhances phagocytosis	Ma et al. 2004.
2.2	R-type	Mytilus trossulus mussel	Timulate the production of pro-inflammatory cytokines such as TNF-α and IFN-γ	Chikalovets et al. 2013.
2.3	Metal independent galectins	Hemolymph	Major role in B lymphocyte activation	Fredrick and Ravichandran 2012.
3.	Taurine	Animal tissue, marine clam, and other marine organisms	Role in innate immunity by phagocytosis mediated inflammatory response at the site of infection, cytoprotective	Marcinkiewicz and Kontny 2014, Schuller-Levis and Park 2004, M. K. Song et al. 2011.
Immunomodulatory and Antimicrobial Peptides				
1.	Callinectin	Callinectes sapidus, Mytilus galloprovincialis	Antibacterial activity against gram-negative bacteria	Noga et al. 2011, Li et al. 2010.

S.no.	Class	Source	Activity	References
2.1	Clavanin A	*Styela clava*	Activity against fungi, MDR Gram-positive, and Gram-negative bacteria in both *in vitro* and *in vivo* studies	Hee Lee, Cho, and Lehrer 1997.
2.2	Clavanin-MO	A synthetic derivative of Clavanin	Antibacterial, antibiofilm activity against Gram-positive and Gram-negative bacteria also shows anti-inflammatory response	Silva et al. 2016.
3.	Crustins	*C. Maenas, Penaeus monodon, Scylla paramamosain,* and *Pacifastacus leniusculus*	Innate immunity against marine Gram-positive bacteria *Corynebacterium glutamicum*	Smith et al. 2008, Donpudsa et al. 2010, Suleiman, Smith, and Dyrynda 2017, Imjongjirak et al. 2009.
4.	Defensins	*Crassostrea virginica* acidifies gill extract, oysters (*C. Gigas* and *C. Virginica*), and mussels (*Mytilus edulis* and *M. Galloprovincialis*)	The microbial pathogen cell membrane and modulating innate and adaptive immune response	Yang et al. 2004a; 2004b, Zhu and Gao 2013a, Hubert 1996a, Seo et al. 2005.
5.	Myticin	Mediterranean *galloprovincialis*	Activity against Gram-positive bacteria, Gram-negative bacteria	Mitta et al. 1999, Balseiro et al. 2011, Domeneghetti et al. 2015.
6.	Mytilin	*Mytilus galloprovincialis* and *M. Edulis.*	Antimicrobial activity and enhances phagocytosis to protect the host from the spread of infection	Charlet et al. 1996a, Mitta et al. 2000
7.	Mytimycin		Circulating hemocytes and shows antimicrobial activity	Hubert 1996b. Charlet et al. 1996b.

Table 1. (Continued)

Protein Hydrolysates

1.	Chlorella	*Chlorella vulgaris*	Supports innate and specific immune response to recover leukocyte count in peripheral blood and bone marrow cellularity	Morris et al. 2007.
2.	Ecklonia	Alariaceae *Ecklonia cava*	Suppresses the immune response by enhancing the production of anti-inflammatory cytokine	Ahn et al. 2008.
3.	Porphyra	Hydrolysis of algae *Porphyra columbina* with a combination of trypsin and Alcalase.	Suppresses the immune response by enhancing the production of anti-inflammatory cytokine	Cian, Martinez-Augustin, and Drago 2012.
4.	Porphyra columbina	Hydrolysis of *Porphyra columbina* with flavourzyme	Immunomodulatory effect on T lymphocytes, macrophages, and splenocytes.	Cian et al. 2012
5.	Edible red algae	Hydrolysis of Porphyra tenera with a combination of proteases and carbohydrates	Antioxidant and anti-inflammatory response by inhibiting LPS	Vo et al. 2014.
6.	Edible microalgae spirulina	Hydrolysis of filamentous blue-green algae *Spirulina maxima* with pepsin, trypsin, and α-chymotrypsin	Mast cell degranulation, decreasing histamine release	Cai et al. 2013.
7.	Oyster peptide-based enteral nutrition formula	By hydrolysis of oyster *Crassostrea hongkongensis* with pepsin, trypsin, and bromelain	Lymphocyte proliferation and NK cell activation	Y. K. Wang et al. 2010b.

| 8. | Oyster | Hydrolysis of oyster *C. Gigas* with proyease | Anticancer and immunomodulatory potentials | Y. K. Wang et al. 2010b. |
| 9. | Paphia undulata meat | Hydrolysis of Chinese clam *Paphia undulate* with alkaline protease | Enhances lymphocytes proliferation | He, X.Q 2015. |

4.2.3.8. Oysters

Oyster protein hydrolysate has prepared by hydrolysis of oyster *C. gigas* with proyease extracted from *Bacillus* sp. SM98011. BALB/c mice with sarcoma-S180 treated with Oyster protein hydrolysate showed an inhibitory effect in a dose-dependent manner and body weight reduction was observed. Oral administered of oyster protein hydrolysate enhanced macrophage growth, the proliferation of lymphocytes, activation of NK cells in S180 mice. Oyster protein hydrolysate showed anticancer and immunomodulatory potentials in mice (Y. K. Wang et al. 2010b).

4.2.3.9. Paphia Undulata Meat

Paphia undulata meat protein hydrolysate has prepared by hydrolysis of Chinese clam *Paphia undulate* with alkaline protease extracted from *Bacillus subtilis*. *P. undulata* meat protein hydrolysate enhances lymphocytes proliferation and its activity (He, X.Q 2015). Summary of compounds and its activity is listed in Table 1.

5. MEDICINAL PLANTS AND ESKAPE PATHOGENS – WITH SPECIAL REFERENCE TO *KLEBSIELLA PNEUMONIAE*

Antibiotic resistance is an ever growing problem for all bacteria in the ESKAPE category and one of the very crucial bacteria of great medical concern is *Klebsiella pneumoniae* (Effah et al. 2020). It is considered as an opportunistic pathogen causing a wide variety of diseases (Cheepurupalli et al. 2017) and is responsible for one-thirds of all Gram-negative infections affecting humans (Navon-Venezia, Kondratyeva, and Carattoli 2017). One of the major consequences of *Klebsiella* infection is the long duration of hospitalization and high morbidity (Giske et al. 2008)..There are plenty of studies on the emergence of MDR and extremely drug-resistant (XDR) forms of this bacteria and this is a major concern for researchers and clinicians worldwide. With the lack of novel antibiotics, the major thrust is being applied towards the role of plant-based polyherbal formulations not only for controlling *Klebsiella* infections but also for overcoming antibiotic resistance by ESKAPE pathogens. This part of the chapter will primarily focus on studies that have investigated plant-based compounds for their efficacy against ESKAPE pathogens with special reference to their anti-*Klebsiella* activity. To be precise, this review will only focus on that scientifically valid literature that has analysed herbs routinely used in various folklore/tribal medicine and established Indian systems of medicine, to be relevant to the Indian context, based on their regular usage and application.

5.1. Glycosides, Flavonoids and Saponins

Butea superba is a vine whose tubers are known to possess pharmacological properties. isolated a bio-active flavonol glycoside from the stems of this plant and tested it against both fungal and bacterial pathogens, including *Klebsiella* (Panghal, Kaushal, and Yadav 2011). The glycoside 3, 5, 7, 3′, 4′-pentahydroxy-8-methoxy-flavonol-3-O-beta-D-xylopyranozyl (1→2)-alpha-L-rhamnopyranoside, was found to be very effective against *Klebsiella*. A screening of 34 plant species that were part of tribal medicine in the Western Ghats of India, was performed against *Escherichia coli*, *Klebsiella aerogenes*, *Proteus vulgaris* and *Pseudomonas aerogenes* (Perumal Samy, Ignacimuthu, and Sen 1998). The study demonstrated that 16 plants showed significant activity, with potent antibacterial activity recorded for *Cassia fistula*, *Terminalia arjuna*, and *Vitex negundo*, suggesting the authenticity of folk medicine. Subsequently the same group reported the efficacy of *Tridax procumbens*, *Cleome viscosa*, *Acalypha indica* and *Boerhaaria erecta* for their antibacterial activity against both Gram- negative and positive bacteria including *Klebsiella* (Perumal Samy, Ignacimuthu, and Raja 1999). Triterpenoid saponin isolated from the leaves of *Lepidagathis hyalina*, common in northern parts of India showed antimicrobial activity against both pathogenic bacteria and fungi (Yadava 2001). Cardiac glycoside is a poison that is produced abundantly in the plants *Calotropis procera* and *C. gigantea*. In fact, the milky sap produced by these plants is used as an arrow poison. Interestingly, the milky sap rich in alkaloids is directly applied over small wounds or bruises in rural India and this is due to their potent antibacterial activity against a wide range of pathogenic bacteria including *K. pneumoniae* (Pattnaik *et al.*, 2017). Flavonoids and cardiac glycosides have wide ranging pharmaceutical properties and are abundantly found in a variety of plants. In a study on *Gymnema sylvestre,* commonly called as 'Gurmar' and primarily used against hyperglycemia, the leaf flavonoids and glycosides were found to possess potent antibacterial activity against *K. pneumoniae* (Arora and Sood 2017) with the added benefits of they being non-toxic. Thus, it is clear that primarily antioxidative flavonoids are equally antibacterial as has been shown for flavonoids from *Albizia odoratissima,* a top nitrogen fixing tree and commonly known as black siris (Banothu et al. 2017). Three of the very common ingredients of Indian cuisine and medicine, *Curcuma longa*, *Zingiber officinale* and *Tinospora cordifolia* are very well known world over. These plants are endowed with a rich variety of phytochemicals and widespread use of these plants is well justified by studies showing their potential against MDR strains of pathogenic bacteria including *K. pneumoniae*. In fact the results obtained by (Chakraborty *et al.* 2014) clearly show that these plants are even better than standard antibiotics, as they were very effective at lower MIC and MBC.

5.2. Tannins and Gallotannins

Tannins are astringent polyphenolic biomolecules widely distributed in many plants, and considered as a major plant defence molecule. They have been exploited for their antibacterial property in different systems of medicine in India. Tannins from the bark of *Prosopis chilensis*, commonly called as Chilean mesquite; *Pithecellobium dulce* commonly called as Singri; and *Mangifera indica*, commonly called as mango, showed significant antibacterial activity against *K. pneumoniae*, indicating the importance of tannins (M. Singh et al. 2010). *Anogeissus latifolia*, commonly known as axlewood, is a very useful tree species native to India, Nepal, Srilanka and Myanmar. Its leaves are a rich source of gallotannins.

The wound healing properties and antibacterial potential of the ethanolic extracts of bark of this tree were tested and though it was found to be good for wound healing, it was only moderately antibacterial against *Staphylococcus aureus*, *E. coli*, *P. aeruginosa* and *K. pneumoniae*, when compared to the standard drugs erythromycine and tetracycline. Thus it appears that bark of this plant might be beneficial against skin ailments, with mild bacterial killing activity (Govindarajan et al. 2004). Perhaps leaf gallotannin could be more potent in this regard. Another large scale antimicrobial screening study was performed for around 61 species of Indian medicinal plants that are used extensively by indigenous people in their traditional medical practices (V. P. Kumar et al. 2006). The plants screened were distributed across 33 families and were tested against both bacterial and fungal pathogens. The authors used crude extracts of the plants and their results showed that 28 plants were very effective. These plants are *Dorema ammmoniacum*, a desert plant commonly known as gomorresin and native to Asia; *Sphaeranthus indicus*, commonly known as East Indian globe thistle and primarily an anti-inflammatory agent; *Dracaena cinnabari*, commonly known as dragon blood tree, native to Socotra archipelago; *Mallotus philippensis*, commonly known kamala tree or kumkum tree; *Jatropha gossypiifolia*, commonly known as bellyache bush or black physicnut native to South America and India; *Aristolochia indica*, a creeper commonly known as Garudakkodi (well known to be antibacterial: see also (Venkatadri et al. 2015)), is native to Southern India and Srilanka; *Lantana camara*, a perennial shrub and commonly known as big-sage or wild-sage, is actually an invasive species; *Nardostachys jatamansi*, native to the Himalayas and commonly known as muskroot; *Randia dumetorum*, commonly known as indigoberry; *Cassia fistula*, commonly known as golden shower or Indian laburnum, is native to the Indian subcontinent and regions of Southeast Asia.

5.3. Polyherbal Formulation

The effectiveness of extracts of medicinal herbs was also shown to work under clinical settings, when Triphala (a polyherbal formulation containing

Terminalia chebula, T. belevica and *Embilica officinalis*) was tested for its antibacterial activity against bacterial isolates (including *K. pneumoniae*) obtained from HIV positive patients (R. Srikumar et al. 2007). The results clearly showed the bacterial growth inhibitory effects of not only the polyherbal Triphala but also its components in the following order of effectiveness: *E. officinalis* > *T. belerica* > *T. chebula*. Thus, it is clear that polyherbals could essentially represent the total of their components, making them effective antibacterial agents.

5.4. Mode of Application of Medicinal Herbs

The ways in which these medicinal plants have been applied for deriving their antibacterial effects are also very interesting. One of the 'Hindu' practices is fire worship and it is usually the norm that only a certain type of wood is used along with other medicinal herbs. It is well known that these plants are also an integral part of the Indian systems of medicine. Thus, the study by (Nautiyal, Chauhan, and Nene 2007) has clearly shown that smoke emanating from burning a particular kind of wood along with other odoriferous and medicinal herbs, was able to reduce the aerial count of human pathogenic bacteria (including *Klebsiella* sp.) in a closed room by over 94% and what was interesting was that the count remained low for up to 30 days (open room) after the treatment, suggesting the antibacterial potential of these traditional medicinal herbs. Such studies do have support from other findings on similar lines (Mohagheghzadeh et al. 2006; Braithwaite, van Vuuren, and Viljoen 2008)

5.5. Effectiveness against Human Isolates of Pathogens

The validity of these medicinal plants against MDR strains of *K. pneumoniae* has also been demonstrated by the findings of Khan *et al.* (2009). Their study used both MDR and ATCC strains of *K. pneumoniae*. Interestingly, the MDR strain of *K. pneumoniae* showed susceptibility to

Acacia nilotica, *Syzgium aromaticum* and *Cinnamum zeylanicum*, while they were strongly resistant to *Terminalia arjuna* and *Eucalyptus globulus*. Their findings showed that *A. nilotica*, commonly known as the gum Arabic tree, was the most potent. *K. pneumoniae* is a major urinary tract pathogen and it is quite difficult to treat it using standard antibiotics, if the strain is a resistant one. A study by (Sharma et al. 2009) that used MDR isolates from urinary tract infections showed that ethanol extracts of *Terminalia chebula*, commonly known as black- or chebulic myrobelan and *Ocimum sanctum*, commonly known as tulsi, were very effective against *K. pneumoniae*. *O. sanctum* is a very sacred plant in Hindu traditions and is a major component of various Indian systems of medicine. Cancer patients and other immune compromised individuals are known to suffer from opportunistic bacterial infections. One of the well recognized opportunistic pathogen is *K. pneumoniae*. It is known to produce very severe infections in these patients. Owing to antibiotic resistance, it is interesting to note that medicinal herbs can control these secondary infections. In a study by (Panghal, Kaushal, and Yadav 2011), the authors demonstrated just that by testing 10 medicinal plants against clinical isolates of different bacteria and fungi obtained from oral cancer patients. Their results conclusively show that eight of the medicinal plants (*Asphodelus tensifolius*, *Asparagus racemosus*, *Balanites aegyptiaca*, *Eclipta alba*, *Murraya koenigii*, *Pedalium murex*, *Ricinus communis* and *Trigonella foenum graecum*) showed very significant antibacterial activity. These results indicate that not only cancer patients but cancer survivors could also be treated with these medicinal herbs to prevent secondary infections by opportunistic pathogens. This is also necessary given that most of these medicinal herbs produce no or very minimal side effects. One of the very famous and very commonly used plants in all branches of Indian system of medicine is *Cassia auriculata*. This plant is commonly known as avaram or avaram senna. Most of the plant has some or other medicinal value. It is then no surprise that oleanolic acid isolated from the leaves of this plant (Senthilkumar and Reetha 2011) was observed to be highly antibacterial, including against *K. pneumoniae*. On similar lines, a coumarin Ulopetrol, isolated from the leaves of *Toddalia asiatica*, commonly known as an orange climber, was for the first time shown to be highly antibacterial against different pathogenic bacteria including *K. pneumoniae* (Karunai Raj et al. 2012) (Rath and Padhy 2015)successfully demonstrated the antibacterial potential of *Holarrhena antidysenterica, Terminalia alata, T. arjuna* and *Paederia foetida*, which were used by an Indian aborigine, against clinical isolates of enteropathogenic bacteria, obtained from hospitalized children.

5.6. A Few Interesting Antibacterial Mechanisms of Medicinal Plants

A 14.3 KDa protease inhibitor was isolated from a commonly used vegetable, *Coccinia grandis*, commonly called as the ivy gourd or kowai fruit. This protease inhibitor showed strong antibacterial activity against pathogenic *K. pneumoniae* (Satheesh and Murugan 2011), suggesting the importance of this plant as part of the regular diet. It is consumed raw as well.

K. pneumoniae is a common causative agent of urinary tract infections and one of the mechanisms that enable successful colonization of the urinary tract is the presence of urease enzyme. Thus, medicinal herbs with potential anti-urease activity could prove to be a useful therapeutic strategy. Indeed, the study by (Bai et al. 2015) where the authors screened for anti-urease activity of Indian medicinal plants revealed the presence of this in *Acacia nilotica, Emblica officinalis, Psidium guajava, Rosa indica* and *Terminalia chebula*, when these plants were tested against different urease producing bacteria.

5.7. Antibiofilm Potential of Medicinal Plants

Medicinal plants have also been shown to be effective against biofilms of pathogenic bacteria. When tested against *K. pneumoniae* biofilms, flavonoids, diterpenes and cardiac glycosides from the barks of *Prunus cerasoides* were found to possess anti-biofilm activity (Arora and Mahajan

2019) suggesting the ability of medicinal herbs in targeting the more antibiotic resistant biofilms of human pathogenic bacteria. Thus, it is clear that medicinal plants are a major source for antibacterial molecules (this knowledge is with us and is also in use since ancient times) and they play a major role in our fight against microbes, especially at a time when most human pathogens have successfully developed resistance against much of the antibiotics currently in use. In fact world over, efforts are being directed towards the identification and development of plant-based bioactive agents that could aid in fighting these microbes (Yu et al. 2020).

5.8. Can Medicinal Plants Prevent the 'Escape' of ESKAPE Pathogens?

In an interesting study by (Chattopadhyay *et al.* 2001), the authors tested methanolic crude and methanol-aqueous extract of leaves and n-butanol crude extract of *Alstonia macrophylla*, commonly known as hard milkwood, for antibacterial activity against a variety of pathogenic bacteria including *Klebsiella* sp. However, *Klebsiella* sp. and *Vibrio cholerae* were found to be resistant to the extracts even up to 2000 µg/ml. This shows that the antibacterial activity of these herbs may not be a universal solution to most bacterial pathogens and indeed certain species such as *Klebsiella* might develop resistance against these plants. It is also important to realize that not all medicinal herbs are equally potent and safe if consumed without proper guidance. Some of these medicinal plants are extremely toxic, while for some, studies have shown them to be effective only against certain pathogens (Sharma et al. 2009) (Sagar and Vidyasagar 2010) (Savithramma et al. 2012). There is an increasing number of studies that are beginning to show that certain bacteria are resistant to some of these medicinal plants. Keeping this in mind it is necessary to also unravel the mechanism of resistance, which in most probability could be similar to that shown for standard antibiotics. Nevertheless, other strategies could overcome the development of resistance against medicinal plants and one such strategy could be the use of nanotechnology.

5.9. Use of Medicinal Plants in Green Nanotechnology

Nanotechnology based application of plant molecules for better antibacterial strategies are gaining traction worldwide. A search of the literature on green synthesis of nanoparticles and their antibacterial potential clearly shows hundreds of research studies that have tried to exploit the antibacterial potential of both the nanoparticle and the plant used for green synthesis. Studies including ours (Manikandan et al. 2015) have shown the efficacy and suitability of a variety of green synthesized nanoparticles for their antibacterial properties ((Prabakar et al. 2013; Ahluwalia et al. 2018; Chandra et al. 2019)Such green synthesized nanoparticles have started to find novel applications in a lot of areas including biomedical sciences (Manjula et al. 2019). What is important to note in all these is the use of medicinal herbs for nanoparticle synthesis. This again is based on the logic that green synthesis would reduce nanoparticle toxicity while increasing the antibacterial efficacy. Thus, it is very apparent that future of antibacterial research is going to be in these lines (Ahmed, Raman, and Veerappan 2016), combining medicinal herbs and nanotechnology, which has a higher probability of yielding results as compared to the often slower and almost complete lack of progress seen in the field of antibiotics.

REFERENCES

Ahlmann, Martina, and Georg Hempel. 2016. "The Effect of Cyclophosphamide on the Immune System: Implications for Clinical Cancer Therapy." *Cancer Chemotherapy and Pharmacology*. doi: 10.1007/s00280-016-3152-1.

Ahluwalia, Vivek, Sasikumar Elumalai, Vinod Kumar, Sandeep Kumar, and Rajender Singh Sangwan. 2018. "Nano Silver Particle Synthesis Using Swertia Paniculata Herbal Extract and Its Antimicrobial Activity." *Microbial Pathogenesis* 114. doi: 10.1016/j.micpath.2017.11.052.

Ahmed, Khan Behlol Ayaz, Thiagarajan Raman, and Anbazhagan Veerappan. 2016. "Future Prospects of Antibacterial Metal

Nanoparticles as Enzyme Inhibitor." *Materials Science and Engineering C.* doi: 10.1016/j.msec.2016.06.034.

Ahn, Ginnae, Insun Hwang, Eunjin Park, Jinhe Kim, You Jin Jeon, Jehee Lee, Jae Woo Park, and Youngheun Jee. 2008. "Immunomodulatory Effects of an Enzymatic Extract from Ecklonia Cava on Murine Splenocytes." *Marine Biotechnology* 10 (3). doi: 10.1007/s10126-007-9062-9.

Alamgir, Mahiuddin, and Shaikh Jamal Uddin. 2010. "Recent Advances on the Ethnomedicinal Plants as Immunomodulatory Agents." *Ethnomedicine: A Source of Complementary Therapeutics* 661 (2).

Arancibia, Sergio, Cecilia Espinoza, Fabián Salazar, Miguel del Campo, Ricardo Tampe, Ta Ying Zhong, Pablo de Ioannes, et al. 2014. "A Novel Immunomodulatory Hemocyanin from the Limpet Fissurella Latimarginata Promotes Potent Anti-Tumor Activity in Melanoma." *PLoS ONE* 9 (1). doi: 10.1371/journal.pone.0087240.

Arora, Daljit Singh, and Himadri Mahajan. 2019. "Major Phytoconstituents of Prunus Cerasoides Responsible for Antimicrobial and Antibiofilm Potential Against Some Reference Strains of Pathogenic Bacteria and Clinical Isolates of MRSA." *Applied Biochemistry and Biotechnology* 188 (4). doi: 10.1007/s12010-019-02985-4.

Arora, Daljit Singh, and Henna Sood. 2017. "In Vitro Antimicrobial Potential of Extracts and Phytoconstituents from Gymnema Sylvestre R.Br. Leaves and Their Biosafety Evaluation." *AMB Express* 7 (1). doi: 10.1186/s13568-017-0416-z.

Bai, Sheema, Pooja Bharti, Leena Seasotiya, Anupma Malik, and Sunita Dalal. 2015. "In Vitro Screening and Evaluation of Some Indian Medicinal Plants for Their Potential to Inhibit Jack Bean and Bacterial Ureases Causing Urinary Infections." *Pharmaceutical Biology* 53 (3). doi: 10.3109/13880209.2014.918158.

Balakrishnan, Bijinu, Binod Prasad, Amit Kumar Rai, Suresh Puthanveetil Velappan, Mahendrakar Namadev Subbanna, and Bhaskar Narayan. 2011. "In Vitro Antioxidant and Antibacterial Properties of Hydrolysed Proteins of Delimed Tannery Fleshings: Comparison of Acid Hydrolysis and Fermentation Methods." *Biodegradation* 22 (2). doi: 10.1007/ s10532-010-9398-0.

Balseiro, Pablo, Alberto Falcó, Alejandro Romero, Sonia Dios, Alicia Martínez-López, Antonio Figueras, Amparo Estepa, and Beatriz Novoa. 2011. "Mytilus Galloprovincialis Myticin C: A Chemotactic Molecule with Antiviral Activity and Immunoregulatory Properties." *PLoS ONE* 6 (8). doi: 10.1371/journal.pone.0023140.

Banothu, Venkanna, Chandrasekharnath Neelagiri, Uma Adepally, Jayalakshmi Lingam, and Kesavaharshini Bommareddy. 2017. "Phytochemical Screening and Evaluation of in Vitro Antioxidant and Antimicrobial Activities of the Indigenous Medicinal Plant Albizia Odoratissima." *Pharmaceutical Biology* 55 (1). doi: 10.1080/13880209. 2017.1291694.

Beaulieu, Lucie, Jacinthe Thibodeau, Claudie Bonnet, Piotr Bryl, and Marie Élise Carbonneau. 2013. "Detection of Antibacterial Activity in an Enzymatic Hydrolysate Fraction Obtained from Processing of Atlantic Rock Crab (Cancer Irroratus) by-Products." *PharmaNutrition* 1 (4). doi: 10.1016/j.phanu.2013.05.004.

Bharani, Shendige Eswara Rao, Mohammed Asad, Sunil Samson Dhamanigi, and Gowda Kallenahalli Chandrakala. 2010. "Immunomodulatory Activity of Methanolic Extract of Morus Alba Linn. (Mulberry) Leaves." *Pakistan Journal of Pharmaceutical Sciences* 23 (1).

Braithwaite, M., S. F. van Vuuren, and A. M. Viljoen. 2008. "Validation of Smoke Inhalation Therapy to Treat Microbial Infections." *Journal of Ethnopharmacology* 119 (3). doi: 10.1016/j.jep.2008.07.050.

Burns, Tamika, Maria Abadi, and Liise Anne Pirofski. 2005. "Modulation of the Lung Inflammatory Response to Serotype 8 Pneumococcal Infection by a Human Immunoglobulin M Monoclonal Antibody to Serotype 8 Capsular Polysaccharide." *Infection and Immunity* 73 (8). doi: 10.1128/IAI.73.8.4530-4538.2005.

Cai, Bingna, Jianyu Pan, Yuantao Wu, Peng Wan, and Huili Sun. 2013. "Immune Functional Impacts of Oyster Peptide-Based Enteral Nutrition

Formula (OPENF) on Mice: A Pilot Study." *Chinese Journal of Oceanology and Limnology* 31 (4). doi: 10.1007/s00343-013-2311-z.

Campo, Miguel del, Sergio Arancibia, Esteban Nova, Fabián Salazar, Andrea González, Bruno Moltedo, Pablo de Ioannes, Jorge Ferreira, Augusto Manubens, and María Inés Becker. 2011. "[Hemocyanins as Immunostimulants]." *Revista Medica de Chile* 139 (2). doi: /S0034-98872011000200015.

Casadevall, A. 2006. "The Third Age of Antimicrobial Therapy." *Clinical Infectious Diseases* 42 (10). doi: 10.1086/503431.

Casadevall, A., and M. D. Scharff. 1994. "Serum Therapy Revisited: Animal Models of Infection and Development of Passive Antibody Therapy." *Antimicrobial Agents and Chemotherapy*. doi: 10.1128/AAC.38.8.1695.

Casadevall, Arturo, and Liise Anne Pirofski. 2004. "New Concepts in Antibody-Mediated Immunity." *Infection and Immunity*. doi: 10.1128/IAI.72.11.6191-6196.2004.

Casadevall, Arturo, and Liise anne Pirofski. 2006. "A Reappraisal of Humoral Immunity Based on Mechanisms of Antibody-Mediated Protection Against Intracellular Pathogens." *Advances in Immunology*. doi: 10.1016/S0065-2776(06)91001-3.

Catanzaro, Michele, Emanuela Corsini, Michela Rosini, Marco Racchi, and Cristina Lanni. 2018. "Immunomodulators Inspired by Nature: A Review on Curcumin and Echinacea." *Molecules*. doi: 10.3390/molecules23112778.

Chalamaiah, Meram, Wenlin Yu, and Jianping Wu. 2018. "Immunomodulatory and Anticancer Protein Hydrolysates (Peptides) from Food Proteins: A Review." *Food Chemistry*. doi: 10.1016/j.foodchem.2017.10.087.

Chandra, Harish, Deepak Patel, Pragati Kumari, J. S. Jangwan, and Saurabh Yadav. 2019. "Phyto-Mediated Synthesis of Zinc Oxide Nanoparticles of Berberis Aristata: Characterization, Antioxidant Activity and Antibacterial Activity with Special Reference to Urinary Tract Pathogens." *Materials Science and Engineering C* 102. doi: 10.1016/j.msec.2019.04.035.

Charlet, Maurice, Serguey Chernysh, Hervé Philippe, Charles Hetru, Jules A. Hoffmann, and Philippe Bulet. 1996a. "Innate Immunity: Isolation of Several Cysteine-Rich Antimicrobial Peptides from the Blood of a Mollusc, Mytilus Edulis." *Journal of Biological Chemistry* 271 (36). doi: 10.1074/jbc.271.36.21808.

———. 1996b. "Innate Immunity: Isolation of Several Cysteine-Rich Antimicrobial Peptides from the Blood of a Mollusc, Mytilus Edulis." *Journal of Biological Chemistry* 271 (36). doi: 10.1074/jbc.271.36.21808.

Cheepurupalli, Lalitha, Thiagarajan Raman, Sudarshan S. Rathore, and Jayapradha Ramakrishnan. 2017. "Bioactive Molecule from Streptomyces Sp. Mitigates MDR Klebsiella Pneumoniae in Zebrafish Infection Model." *Frontiers in Microbiology* 8 (APR). Frontiers Research Foundation. doi: 10.3389/fmicb.2017.00614.

Chikalovets, I. v., A. S. Kondrashina, O. v. Chernikov, V. I. Molchanova, and P. A. Luk'Yanov. 2013. "Isolation and General Characteristics of Lectin from the Mussel Mytilus Trossulus." *Chemistry of Natural Compounds* 48 (6). doi: 10.1007/s10600-013-0463-x.

Cian, Raúl E., Rocío López-Posadas, Silvina R. Drago, Fermín Sánchez De Medina, and Olga Martínez-Augustin. 2012. "A Porphyra Columbina Hydrolysate Upregulates IL-10 Production in Rat Macrophages and Lymphocytes through an NF-KB, and P38 and JNK Dependent Mechanism." *Food Chemistry* 134 (4). doi: 10.1016/j.foodchem.2012.03.134.

Cian, Raúl E., Olga Martínez-Augustin, and Silvina R. Drago. 2012. "Bioactive Properties of Peptides Obtained by Enzymatic Hydrolysis from Protein Byproducts of Porphyra Columbina." *Food Research International* 49 (1). doi: 10.1016/j.foodres.2012.07.003.

Cornejo-García, José A., James R. Perkins, Raquel Jurado-Escobar, Elena García-Martín, José A. Agúndez, Enrique Viguera, Natalia Pérez-Sánchez, and Natalia Blanca-López. 2016. "Pharmacogenomics of Prostaglandin and Leukotriene Receptors." *Frontiers in Pharmacology*. doi: 10.3389/fphar.2016.00316.

Dolashka-Angelova, Pavlina, Tsetanka Stefanova, Evangelia Livaniou, Lyudmila Velkova, Persefoni Klimentzou, Stefan Stevanovic, B. Salvato, Hristo Neychev, and Wolfgang Voelter. 2008. "Immunological Potential of Helix Vulgaris and Rapana Venosa Hemocyanins." *Immunological Investigations* 37 (8). doi: 10.1080/08820130802403366.

Domeneghetti, Stefania, Marco Franzoi, Nunzio Damiano, Rosa Norante, Nancy M. El Halfawy, Stefano Mammi, Oriano Marin, Massimo Bellanda, and Paola Venier. 2015. "Structural and Antimicrobial Features of Peptides Related to Myticin C, a Special Defense Molecule from the Mediterranean Mussel Mytilus Galloprovincialis." *Journal of Agricultural and Food Chemistry* 63 (42). doi: 10.1021/acs.jafc.5b03491.

Donpudsa, Suchao, Vichien Rimphanitchayakit, Anchalee Tassanakajon, Irene Söderhäll, and Kenneth Söderhäll. 2010. "Characterization of Two Crustin Antimicrobial Peptides from the Freshwater Crayfish Pacifastacus Leniusculus." *Journal of Invertebrate Pathology* 104 (3). doi: 10.1016/j.jip.2010.04.001.

Effah, Clement Yaw, Tongwen Sun, Shaohua Liu, and Yongjun Wu. 2020. "Klebsiella Pneumoniae: An Increasing Threat to Public Health." *Annals of Clinical Microbiology and Antimicrobials*. doi: 10.1186/s12941-019-0343-8.

Ekins, Sean, Mary A. Lingerfelt, Jason E. Comer, Alexander N. Freiberg, Jon C. Mirsalis, Kathleen O'Loughlin, Anush Harutyunyan, Claire McFarlane, Carol E. Green, and Peter B. Madrid. 2018. "Efficacy of Tilorone Dihydrochloride against Ebola Virus Infection." *Antimicrobial Agents and Chemotherapy* 62 (2). doi: 10.1128/AAC.01711-17.

Flores, Camila, Guillemette Fouquet, Ivan Cruz Moura, Thiago Trovati Maciel, and Olivier Hermine. 2019. "Lessons to Learn from Low-Dose Cyclosporin-A: A New Approach for Unexpected Clinical Applications." *Frontiers in Immunology*. doi: 10.3389/fimmu.2019.00588.

Fredrick, W. Sylvester, and S. Ravichandran. 2012. "Hemolymph Proteins in Marine Crustaceans." *Asian Pacific Journal of Tropical Biomedicine* 2 (6). doi: 10.1016/S2221-1691(12)60084-7.

Ghaisas, M. M., S. A. Shaikh, and A. D. Deshpande. 2009. "Evaluation of the Immunomodulatory Activity of Ethanolic Extract of the Stem Bark of Bauhinia Variegata Linn." *International Journal of Green Pharmacy* 3 (1). doi: 10.4103/0973-8258.49379.

Giske, Christian G., Dominique L. Monnet, Otto Cars, and Yehuda Carmeli. 2008. "Clinical and Economic Impact of Common Multidrug-Resistant Gram-Negative Bacilli." *Antimicrobial Agents and Chemotherapy*. doi: 10.1128/AAC.01169-07.

Gong, Yan, Sarah Louise Klingenberg, and Christian Gluud. 2004. "D-Penicillamine for Primary Biliary Cirrhosis." *Cochrane Database of Systematic Reviews*. doi: 10.1002/14651858.cd004789.pub2.

Govindarajan, Raghavan, Madhavan Vijayakumar, Chandana Venkateshwara Rao, Annie Shirwaikar, Shanta Mehrotra, and Palpu Pushpangadan. 2004. "Healing Potential of Anogeissus Latifolia for Dermal Wounds in Rats." *Acta Pharmaceutica* 54 (4).

Gruppen, Mariken P., Antonia H. Bouts, Marijke C. Jansen-van der Weide, Maruschka P. Merkus, Aleksandra Zurowska, Michal Maternik, Laura Massella, et al. 2018. "A Randomized Clinical Trial Indicates That Levamisole Increases the Time to Relapse in Children with Steroid-Sensitive Idiopathic Nephrotic Syndrome." *Kidney International* 93 (2). doi: 10.1016/j.kint.2017.08.011.

Harborne, Jeffrey B. 1987. "Economic and Medicinal Plant Research, Volume 1." *Phytochemistry* 26 (5). doi: 10.1016/s0031-9422(00)81867-1.

Havla, J., T. Kümpfel, and R. Hohlfeld. 2015. "Immuntherapie Der Multiplen Sklerose: Überblick Und Update [Multiple Sclerosis Immunotherapy: Overview And Update]." *Internist* 56 (4). doi: 10.1007/s00108-015-3668-1.

Hee Lee, I. N., Yoon Cho, and Robert I. Lehrer. 1997. "Effects of PH and Salinity on the Antimicrobial Properties of Clavanins." *Infection and Immunity* 65 (7). doi: 10.1128/iai.65.7.2898-2903.1997.

Hubert, Florence. 1996a. "A Member of the Arthropod Defensin Family from Edible Mediterranean Mussels (Mytilus Galloprovincialis)." *European Journal of Biochemistry* 240 (1). doi: 10.1111/j.1432-1033.1996.0302h.x.

———. 1996b. "A Member of the Arthropod Defensin Family from Edible Mediterranean Mussels (Mytilus Galloprovincialis)." *European Journal of Biochemistry* 240 (1). doi: 10.1111/j.1432-1033.1996.0302h.x.

Imjongjirak, Chanprapa, Piti Amparyup, Anchalee Tassanakajon, and Siriporn Sittipraneed. 2009. "Molecular Cloning and Characterization of Crustin from Mud Crab Scylla Paramamosain." *Molecular Biology Reports* 36 (5). doi: 10.1007/s11033-008-9253-0.

Indian Medicinal Plants. 2007. *Indian Medicinal Plants*. doi: 10.1007/978-0-387-70638-2.

Jantan, Ibrahim, Waqas Ahmad, and Syed Nasir Abbas Bukhari. 2015. "Plant-Derived Immunomodulators: An Insight on Their Preclinical Evaluation and Clinical Trials." *Frontiers in Plant Science*. doi: 10.3389/fpls.2015.00655.

Kang, Hee Kyoung, Chang Ho Seo, and Yoonkyung Park. 2015. "Marine Peptides and Their Anti-Infective Activities." *Marine Drugs*. doi: 10.3390/md13010618.

Karunai Raj, M., C. Balachandran, V. Duraipandiyan, P. Agastian, and S. Ignacimuthu. 2012. "Antimicrobial Activity of Ulopterol Isolated from Toddalia Asiatica (L.) Lam.: A Traditional Medicinal Plant." *Journal of Ethnopharmacology* 140 (1). doi: 10.1016/j.jep.2012.01.005.

Khan, Vasim, Abul Kalam Najmi, Mohd Akhtar, Mohd Aqil, Mohd Mujeeb, and K K Pillai. 2012. "Effects of Ocimum Basi... [Rev Med Chir Soc Med Nat Iasi. 2007 Oct-Dec] - PubMed - NCBI." *Journal of Pharmacy & Bioallied Sciences*. doi: 10.4103/0975-7406.92727.

Kiewiet, Mensiena B.G., Marijke M. Faas, and Paul de Vos. 2018. "Immunomodulatory Protein Hydrolysates and Their Application." *Nutrients*. doi: 10.3390/nu10070904.

Kumar, Dinesh, Vikrant Arya, Ranjeet Kaur, Zulfiqar Ali Bhat, Vivek Kumar Gupta, and Vijender Kumar. 2012. "A Review of Immunomodulators in the Indian Traditional Health Care System."

Journal of Microbiology, Immunology and Infection. doi: 10.1016/j. jmii.2011.09.030.

Kumar, V. Prashanth, Neelam S. Chauhan, Harish Padh, and M. Rajani. 2006. "Search for Antibacterial and Antifungal Agents from Selected Indian Medicinal Plants." *Journal of Ethnopharmacology* 107 (2). doi: 10.1016/j.jep.2006.03.013.

Lammers, Rianne J.M., Wim P.J. Witjes, Maria H.D. Janzing-Pastors, Christien T.M. Caris, and J. Alfred Witjes. 2012. "Intracutaneous and Intravesical Immunotherapy with Keyhole Limpet Hemocyanin Compared with Intravesical Mitomycin in Patients with Non-Muscle-Invasive Bladder Cancer: Results from a Prospective Randomized Phase III Trial." *Journal of Clinical Oncology* 30 (18). doi: 10.1200/JCO. 2011.39.2936.

Li, Chun, Tor Haug, Morten K. Moe, Olaf B. Styrvold, and Klara Stensvåg. 2010. "Centrocins: Isolation and Characterization of Novel Dimeric Antimicrobial Peptides from the Green Sea Urchin, Strongylocentrotus Droebachiensis." *Developmental and Comparative Immunology* 34 (9). doi: 10.1016/j.dci.2010.04.004.

Long, Valencia. 2016. "Aloe Vera in Dermatology-the Plant of Immortality." *JAMA Dermatology.* doi: 10.1001/jamadermatol.2016. 0077.

Ma, Young Gerl, Mi Yhang Cho, Mingyi Zhao, Ji Won Park, Misao Matsushita, Teizo Fujita, and Bok Luel Lee. 2004. "Human Mannose-Binding Lectin and L-Ficolin Function as Specific Pattern Recognition Proteins in the Lectin Activation Pathway of Complement." *Journal of Biological Chemistry* 279 (24). doi: 10.1074/jbc.M400701200.

Manikandan, Ramar, Beulaja Manikandan, Thiagarajan Raman, Koodalingam Arunagirinathan, Narayanan Marimuthu Prabhu, Muthuramalingam Jothi Basu, Muthulakshmi Perumal, Subramanian Palanisamy, and Arumugam Munusamy. 2015. "Biosynthesis of Silver Nanoparticles Using Ethanolic Petals Extract of Rosa Indica and Characterization of Its Antibacterial, Anticancer and Anti-Inflammatory Activities." *Spectrochimica Acta - Part A: Molecular and Biomolecular Spectroscopy* 138. doi: 10.1016/j.saa.2014.10.043.

Manjula, Selvadurai, Olapalayam Lakshmanan Shanmugasundaram, Balasubramanian Mythili Gnanamangai, Ramalingam Pavithra, Shivaji Kavitha, and Ponnusamy Ponmurugan. 2019. "Plasma Treated Fabrics Coated with Naturally Derived Ag-NPs for Biomedical Application." *IET Nanobiotechnology* 13 (4). doi: 10.1049/iet-nbt.2018.5218.

Manu, Kanjoormana Aryan, and Girija Kuttan. 2009. "Immunomodulatory Activities of Punarnavine, an Alkaloid from Boerhaavia Diffusa." *Immunopharmacology and Immunotoxicology* 31 (3). doi: 10.1080/08923970802702036.

Marcinkiewicz, Janusz, and Ewa Kontny. 2014. "Taurine and Inflammatory Diseases." *Amino Acids*. doi: 10.1007/s00726-012-1361-4.

McFadden, David W., Dale R. Riggs, Barbara J. Jackson, and Linda Vona-Davis. 2003. "Keyhole Limpet Hemocyanin, a Novel Immune Stimulant with Promising Anticancer Activity in Barrett's Esophageal Adenocarcinoma." *American Journal of Surgery* 186 (5). doi: 10.1016/j.amjsurg.2003.08.002.

Miccadei, Stefania, Roberta Masella, Anna Maria Mileo, and Sandra Gessani. 2016. "Ω3 Polyunsaturated Fatty Acids as Immunomodu-lators in Colorectal Cancer: New Potential Role in Adjuvant Therapies." *Frontiers in Immunology*. doi: 10.3389/fimmu.2016.00486.

Mitta, Guillaume, Florence Hubert, Elisabeth A. Dyrynda, Pierre Boudry, and Philippe Roch. 2000. "Mytilin B and MGD2, Two Antimicrobial Peptides of Marine Mussels: Gene Structure and Expression Analysis." *Developmental and Comparative Immunology* 24 (4). doi: 10.1016/S0145-305X(99)00084-1.

Mitta, Guillaume, Florence Hubert, Thierry Noël, and Philippe Roch. 1999. "Myticin, a Novel Cysteine-Rich Antimicrobial Peptide Isolated from Haemocytes and Plasma of the Mussel Mytilus Galloprovincialis." *European Journal of Biochemistry* 265 (1). doi: 10.1046/j.1432-1327.1999.00654.x.

Mohagheghzadeh, Abdolali, Pouya Faridi, Mohammadreza Shams-Ardakani, and Younes Ghasemi. 2006. "Medicinal Smokes." *Journal of Ethnopharmacology*. doi: 10.1016/j.jep.2006.09.005.

Morris, Humberto J., Olimpia Carrillo, Angel Almarales, Rosa C. Bermúdez, Yamila Lebeque, Roberto Fontaine, Gabriel Llauradó, and Yaixa Beltrán. 2007. "Immunostimulant Activity of an Enzymatic Protein Hydrolysate from Green Microalga Chlorella Vulgaris on Undernourished Mice." *Enzyme and Microbial Technology* 40 (3). doi: 10.1016/j.enzmictec.2006.07.021.

Murray, B., and R. FitzGerald. 2007. "Angiotensin Converting Enzyme Inhibitory Peptides Derived from Food Proteins: Biochemistry, Bioactivity and Production." *Current Pharmaceutical Design* 13 (8). doi: 10.2174/138161207780363068.

Naidoo, J., D. B. Page, and J. D. Wolchok. 2014. "Immune Modulation for Cancer Therapy." *British Journal of Cancer*. doi: 10.1038/bjc.2014.348.

Nautiyal, Chandra Shekhar, Puneet Singh Chauhan, and Yeshwant Laxman Nene. 2007. "Medicinal Smoke Reduces Airborne Bacteria." *Journal of Ethnopharmacology* 114 (3). doi: 10.1016/j.jep.2007.08.038.

Navon-Venezia, Shiri, Kira Kondratyeva, and Alessandra Carattoli. 2017. "Klebsiella Pneumoniae: A Major Worldwide Source and Shuttle for Antibiotic Resistance." *FEMS Microbiology Reviews*. doi: 10.1093/femsre/fux013.

Noga, Edward J., Kathryn L. Stone, Abbey Wood, William L. Gordon, and David Robinette. 2011. "Primary Structure and Cellular Localization of Callinectin, an Antimicrobial Peptide from the Blue Crab." *Developmental and Comparative Immunology* 35 (4). doi: 10.1016/j.dci.2010.11.015.

Noori, Shokoofeh, Gholam Ali Naderi, Zuhair M. Hassan, Zohre Habibi, S. Zahra Bathaie, and S. Mahmoud M. Hashemi. 2004. "Immunosuppressive Activity of a Molecule Isolated from Artemisia Annua on DTH Responses Compared with Cyclosporin A." *International Immunopharmacology* 4 (10–11). doi: 10.1016/j.intimp.2004.05.003.

Okolie, Chigozie Louis, Subin R. Subin, Chibuike C. Udenigwe, Alberta N.A. Aryee, and Beth Mason. 2017. "Prospects of Brown Seaweed Polysaccharides (BSP) as Prebiotics and Potential Immunomodulators." *Journal of Food Biochemistry*. doi: 10.1111/jfbc.12392.

Ozkan, Melda Comert, Murat Tombuloglu, Fahri Sahin, and Guray Saydam. 2015. "Evaluation of Immunomodulatory Drugs in Multiple Myeloma: Single Center Experience." *American Journal of Blood Research* 5 (2).

Panghal, Manju, Vivek Kaushal, and Jaya P. Yadav. 2011. "In Vitro Antimicrobial Activity of Ten Medicinal Plants against Clinical Isolates of Oral Cancer Cases." *Annals of Clinical Microbiology and Antimicrobials* 10. doi: 10.1186/1476-0711-10-21.

Perumal Samy, R., S. Ignacimuthu, and D. Patric Raja. 1999. "Preliminary Screening of Ethnomedicinal Plants from India." *Journal of Ethnopharmacology* 66 (2). doi: 10.1016/S0378-8741(99)00038-0.

Perumal Samy, R., S. Ignacimuthu, and A. Sen. 1998. "Screening of 34 Indian Medicinal Plants for Antibacterial Properties." *Journal of Ethnopharmacology* 62 (2). doi: 10.1016/S0378-8741(98)00057-9.

Pirofski, Liise anne, and Arturo Casadevall. 2006. "Immunomodulators as an Antimicrobial Tool." *Current Opinion in Microbiology*. doi: 10.1016/j.mib.2006.08.004.

Pirofski, Liise Anne, and Arturo Casadevall. 2018. "The Damage-Response Framework as a Tool for the Physician-Scientist to Understand the Pathogenesis of Infectious Diseases." *Journal of Infectious Diseases* 218. doi: 10.1093/infdis/jiy083.

Poudel, Santosh, Pradeep, and Mahendra Prasad Yadav. 2019. "Agastya Haritaki Rasayana: A Critical Review." *Journal of Drug Delivery and Therapeutics* 9 (1-s). doi: 10.22270/jddt.v9i1-s.2283.

Prabakar, Kandasamy, Periyasamy Sivalingam, Siyed Ibrahim Mohamed Rabeek, Manickam Muthuselvam, Naresh Devarajan, Annavi Arjunan, Rajamanickam Karthick, Micky Maray Suresh, and John Pote Wembonyama. 2013. "Evaluation of Antibacterial Efficacy of Phyto Fabricated Silver Nanoparticles Using Mukia Scabrella (Musumusukkai) against Drug Resistance Nosocomial Gram Negative Bacterial Pathogens." *Colloids and Surfaces B: Biointerfaces* 104. doi: 10.1016/j.colsurfb.2012.11.041.

Pradhan, D., P. K. Panda, and G. Tripathy. 2009. "Evaluation of the Immunomodulatory Activity of the Methanolic Extract of Couroupita Guianensis Aubl. Flowers in Rats." *Natural Product Radiance* 8 (1).

Raoult, D. 1993. "Antimicrobial Agents and Intracellular Pathogens." *Preferred Customer Sale CRC Press* 4924.

Rath, Shakti, and Rabindra N. Padhy. 2015. "Antibacterial Efficacy of Five Medicinal Plants against Multidrug-Resistant Enteropathogenic Bacteria Infecting under-5 Hospitalized Children." *Journal of Integrative Medicine* 13 (1). doi: 10.1016/S2095-4964(15)60154-6.

Ríos, José Luis. 2010. "Effects of Triterpenes on the Immune System." *Journal of Ethnopharmacology*. doi: 10.1016/j.jep.2009.12.045.

Román, Juan José Mora, Miguel del Campo, Javiera Villar, Francesca Paolini, Gianfranca Curzio, Aldo Venuti, Lilian Jara, et al. 2019. "Immunotherapeutic Potential of Mollusk Hemocyanins in Combination with Human Vaccine Adjuvants in Murine Models of Oral Cancer." *Journal of Immunology Research* 2019. doi: 10.1155/2019/7076942.

Routy, Jean Pierre, Vikram Mehraj, and Wei Cao. 2016. "HIV Immunotherapy Comes of Age: Implications for Prevention, Treatment and Cure." *Expert Review of Clinical Immunology*. doi: 10.1586/1744666X.2016.1112269.

Ruiz-Ruiz, Federico, Elena I. Mancera-Andrade, and Hafiz M. N. Iqbal. 2016. "Marine-Derived Bioactive Peptides for Biomedical Sectors: A Review." *Protein & Peptide Letters* 24 (2). doi: 10.2174/0929866523666160802155347.

Sagar, Kavitha, and G. M. Vidyasagar. 2010. "Antimicrobial Activity of α-(2-Hydroxy-2-Methylpropyl) ω- (2-Hydroxy-3-Methylbut-2-En-1-Yl) Polymethylene from Caesalpinia Bonducella (L.) Flem." *Indian Journal of Pharmaceutical Sciences* 72 (4). doi: 10.4103/0250-474X.73929.

Satheesh, L. Shilpa, and K. Murugan. 2011. "Antimicrobial Activity of Protease Inhibitor from Leaves of Coccinia Grandis (L.) Voigt." *Indian Journal of Experimental Biology* 49 (5).

Savithramma, N., S. Ankanna, M. Linga Rao, and J. Saradvathi. 2012. "Studies on Antimicrobial Efficacy of Medicinal Tuberous Shrub Talinum Cuneifolium." *Journal of Environmental Biology* 33 (4).

Schuller-Levis, Georgia B., and Eunkyue Park. 2004. "Taurine and Its Chloramine: Modulators of Immunity." *Neurochemical Research*. doi: 10.1023/B:NERE.0000010440.37629.17.

Senthilkumar, Palanisamy Kandasamy, and D. Reetha. 2011. "Isolation and Identification of Antibacterial Compound from the Leaves of Cassia Auriculata." *European Review for Medical and Pharmacolo-gical Sciences* 15 (9).

Seo, Jung Kil, J. Myron Crawford, Kathryn L. Stone, and Edward J. Noga. 2005. "Purification of a Novel Arthropod Defensin from the American Oyster, Crassostrea Virginica." *Biochemical and Biophysical Research Communications* 338 (4). doi: 10.1016/j.bbrc.2005.11.013.

Sharififar, Fariba, Shirin Pournourmohammadi, Moslem Arabnejad, Ramin Rastegarianzadeh, Omid Ranjbaran, and Amin Purhemmaty. 2009. "Immunomodulatory Activity of Aqueous Extract of Heracleum Persicum Desf. in Mice." *Iranian Journal of Pharmaceutical Research* 8 (4).

Sharma, Anjana, S. Chandraker, V. K. Patel, and Padmini Ramteke. 2009. "Antibacterial Activity of Medicinal Plants against Pathogens Causing Complicated Urinary Tract Infections." *Indian Journal of Pharmaceutical Sciences* 71 (2). doi: 10.4103/0250-474X.54279.

Sharon, Nathan. 2007. "Lectins: Carbohydrate-Specific Reagents and Biological Recognition Molecules." *The Journal of Biological Chemistry*. doi: 10.1074/JBC.X600004200.

Shrivastava, Neeta, Astha Varma, and Harish Padh. 2011. "Andrographolide: A New Plant-Derived Antineoplastic Entity on Horizon." *Evidence-Based Complementary and Alternative Medicine*. doi: 10.1093/ecam/nep135.

Silva, O. N., C. de La Fuente-Núñez, E. F. Haney, I. C.M. Fensterseifer, S. M. Ribeiro, W. F. Porto, P. Brown, et al. 2016. "An Anti-Infective Synthetic Peptide with Dual Antimicrobial and Immunomodulatory Activities." *Scientific Reports* 6. doi: 10.1038/srep35465.

Singh, Brij Pal, Shilpa Vij, and Subrota Hati. 2014. "Functional Significance of Bioactive Peptides Derived from Soybean." *Peptides*. doi: 10.1016/j.peptides.2014.01.022.

Singh, Meenakshi, Sayyada Khatoon, Shweta Singh, Vivek Kumar, Ajay Kumar Singh Rawat, and Shanta Mehrotra. 2010. "Antimicrobial Screening of Ethnobotanically Important Stem Bark of Medicinal Plants." *Pharmacognosy Research* 2 (4). doi: 10.4103/0974-8490. 69127.

SM, NSNC, Sandeep B Patil, Nilofar S Naikwade, and Chandrakant S Magdum. 2009. "Review on Phytochemistry and Pharmacological Aspects of Euphorbia Hirta Linn." *Jprhc* 1 (1).

Smith, Valerie J., Jorge M.O. Fernandes, Graham D. Kemp, and Chris Hauton. 2008. "Crustins: Enigmatic WAP Domain-Containing Antibacterial Proteins from Crustaceans." *Developmental and Comparative Immunology*. doi: 10.1016/j.dci.2007.12.002.

Song, M. K., N. K. Salam, Basil D. Roufogalis, and T. H.W. Huang. 2011. "Lycium Barbarum (Goji Berry) Extracts and Its Taurine Component Inhibit PPAR-γ-Dependent Gene Transcription in Human Retinal Pigment Epithelial Cells: Possible Implications for Diabetic Retinopathy Treatment." *Biochemical Pharmacology* 82 (9). doi: 10. 1016/j.bcp.2011.07.089.

Song, Ru, Rong Bian Wei, Hong Yu Luo, and Dong Feng Wang. 2012. "Isolation and Characterization of an Antibacterial Peptide Fraction from the Pepsin Hydrolysate of Half-Fin Anchovy (Setipinna Taty)." *Molecules* 17 (3). doi: 10.3390/molecules17032980.

Spellberg, B., J. H. Powers, E. P. Brass, L. G. Miller, and J. E. Edwards. 2004. "Trends in Antimicrobial Drug Development: Implications for the Future." *Clinical Infectious Diseases* 38 (9). doi: 10.1086/420937.

Srikumar, R., N. Jeya Parthasarathy, E. M. Shankar, S. Manikandan, R. Vijayakumar, R. Thangaraj, K. Vijayananth, R. Sheeladevi, and Usha Anand Rao. 2007. "Evaluation of the Growth Inhibitory Activities of Triphala against Common Bacterial Isolates from HIV Infected Patients." *Phytotherapy Research* 21 (5). doi: 10.1002/ptr.2105.

Srikumar, Ramasundaram, Narayanaperumal Jeya Parthasarathy, and Rathinasamy Sheela Devi. 2005. "Immunomodulatory Activity of Triphala on Neutrophil Functions." *Biological and Pharmaceutical Bulletin* 28 (8). doi: 10.1248/bpb.28.1398.

Suleiman, Suzanne, Valerie J. Smith, and Elisabeth A. Dyrynda. 2017. "Unusual Tissue Distribution of Carcinin, an Antibacterial Crustin, in the Crab, Carcinus Maenas, Reveals Its Multi-Functionality." *Developmental and Comparative Immunology* 76. doi: 10.1016/j.dci. 2017.06.010.

Tanaka, Takuji, Haruo Sugiura, Ryoichi Inaba, Akiyoshi Nishikawa, Akira Murakami, Koichi Koshimizu, and Hajime Ohigashi. 1999. "Immunomodulatory Action of Citrus Auraptene on Macrophage Functions and Cytokine Production of Lymphocytes in Female BALB/c Mice." *Carcinogenesis* 20 (8). doi: 10.1093/carcin/20.8.1471.

Tang, Wenting, Hui Zhang, Li Wang, Haifeng Qian, and Xiguang Qi. 2015. "Targeted Separation of Antibacterial Peptide from Protein Hydrolysate of Anchovy Cooking Wastewater by Equilibrium Dialysis." *Food Chemistry* 168. doi: 10.1016/j.foodchem.2014.07.027.

Trabattoni, D., M. Clerici, S. Centanni, M. Mantero, M. Garziano, and F. Blasi. 2017. "Immunomodulatory Effects of Pidotimod in Adults with Community-Acquired Pneumonia Undergoing Standard Antibiotic Therapy." *Pulmonary Pharmacology and Therapeutics* 44. doi: 10. 1016/j.pupt.2017.03.005.

Ulrich, C., J. Bichel, S. Euvrard, B. Guidi, C. M. Proby, P. C.M. van de Kerkhof, P. Amerio, J. Rønnevig, H. B. Slade, and E. Stockfleth. 2007. "Topical Immunomodulation under Systemic Immunosuppression: Results of a Multicentre, Randomized, Placebo-Controlled Safety and Efficacy Study of Imiquimod 5% Cream for the Treatment of Actinic Keratoses in Kidney, Heart, and Liver Transplant Patients." *British Journal of Dermatology* 157 (SUPPL. 2). doi: 10.1111/j.1365-2133. 2007.08269.x.

Venkatadri, B., N. Arunagirinathan, M. R. Rameshkumar, Latha Ramesh, A. Dhanasezhian, and P. Agastian. 2015. "In Vitro Antibacterial Activity of Aqueous and Ethanol Extracts of Aristolochia Indica and Toddalia Asiatica against Multidrug-Resistant Bacteria." *Indian Journal of Pharmaceutical Sciences* 77 (6). doi: 10.4103/0250-474X.174991.

Villani, Alexandra-Chloé, Siranush Sarkizova, and Nir Hacohen. 2018. "Systems Immunology: Learning the Rules of the Immune System."

Annual Review of Immunology 36 (1). doi: 10.1146/annurev-immunol-042617-053035.

Vinothapooshan, G., and K. Sundar. 2011. "Immunomodulatory Activity of Various Extracts of Adhatoda Vasica Linn. in Experimental Rats." *African Journal of Pharmacy and Pharmacology* 5 (3). doi: 10.5897/AJPP10.126.

Vo, Thanh Sang, Dai Hung Ngo, Kyong Hwa Kang, Sun Joo Park, and Se Kwon Kim. 2014. "The Role of Peptides Derived from Spirulina Maxima in Downregulation of FcεRI-Mediated Allergic Responses." *Molecular Nutrition and Food Research* 58 (11). doi: 10.1002/mnfr.201400329.

Wang, Yu Kai, Hai Lun He, Guo Fan Wang, Hao Wu, Bai Cheng Zhou, Xiu Lan Chen, and Yu Zhong Zhang. 2010a. "Oyster (Crassostrea Gigas) Hydrolysates Produced on a Plant Scale Have Antitumor Activity and Immunostimulating Effects in BALB/c Mice." *Marine Drugs* 8 (2). doi: 10.3390/md8020255.

———. 2010b. "Oyster (Crassostrea Gigas) Hydrolysates Produced on a Plant Scale Have Antitumor Activity and Immunostimulating Effects in BALB/c Mice." *Marine Drugs* 8 (2). doi: 10.3390/md8020255.

Wang, Yuefei, Qianfei Huang, Dedong Kong, and Ping Xu. 2018. "Production and Functionality of Food-Derived Bioactive Peptides: A Review." *Mini-Reviews in Medicinal Chemistry* 18 (18). doi: 10.2174/1389557518666180424110754.

Xu, Qiaoqing, Gailing Wang, Hanwen Yuan, Yi Chai, and Zhili Xiao. 2010. "CDNA Sequence and Expression Analysis of an Antimicrobial Peptide, Theromacin, in the Triangle-Shell Pearl Mussel Hyriopsis Cumingii." *Comparative Biochemistry and Physiology - B Biochemistry and Molecular Biology* 157 (1). doi: 10.1016/j.cbpb.2010.05.010.

Yadava, R. N. 2001. "A New Biologically Active Triterpenoid Saponin from the Leaves of Lepidagathis Hyalina Nees." *Natural Product Letters* 15 (5). doi: 10.1080/10575630108041298.

Yang, Rongcun, Francisco Martinez Murillo, Hengmi Cui, Richard Blosser, Satoshi Uematsu, Kiyoshi Takeda, Shizuo Akira, Raphael P. Viscidi, and Richard B. S. Roden. 2004a. "Papillomavirus-Like Particles

Stimulate Murine Bone Marrow-Derived Dendritic Cells To Produce Alpha Interferon and Th1 Immune Responses via MyD88." *Journal of Virology* 78 (20). doi: 10.1128/jvi.78.20.11152-11160.2004.

———. 2004b. "Papillomavirus-Like Particles Stimulate Murine Bone Marrow-Derived Dendritic Cells To Produce Alpha Interferon and Th1 Immune Responses via MyD88." *Journal of Virology* 78 (20). doi: 10.1128/jvi.78.20.11152-11160.2004.

Yu, Zhihui, Jie Tang, Tushar Khare, and Vinay Kumar. 2020. "The Alarming Antimicrobial Resistance in ESKAPEE Pathogens: Can Essential Oils Come to the Rescue?" *Fitoterapia*. Elsevier B.V. doi: 10.1016/j.fitote.2019.104433.

Zhu, S, and B Gao. 2013a. "Evolutionary Origin of Beta-Defensins." *Dev Comp Immunol*, no. 39. doi: S0145-305X(12)00034-1 [pii]\r10.1016/j.dci.2012.02.011 [doi]. 2013b. "Evolutionary Origin of Beta-Defensins." *Dev Comp Immunol*, no. 39.doi:S0145-305X(12)00034-1[pii]\r10.1016/ j.dci.2012.02.011 [doi].

Chapter 2

RECENT ADVANCES IN THE UNDERSTANDING AND MANAGEMENT OF *KLEBSIELLA PNEUMONIAE* INFECTIONS

Muhsin Jamal[1,*]*,
Sayed Muhammad Ata Ullah Shah Bukhari*[1]*,
Sana Raza*[1]*, Liloma Shah*[1]*, Redaina*[1]*, Kanwal Mazhar*[1]
and Saadia Andleeb[2]

[1]Department of Microbiology, Abdul Wali Khan University,
Garden Campus, Mardan, Pakistan
[2]Atta-ur-Rahman School of Applied Biosciences,
National University of Sciences and Technology, Islamabad, Pakistan

ABSTRACT

Klebsiella pneumoniae, Gram-negative bacilli, fits in the family *'Enterobacteriaceae'* and is considered as the normal flora of human beings. It is commonly found in healthcare and community related

---

[*] Corresponding Author's Email: muhsinjamal@awkum.edu.pk.

infections. The frequent occurrence of resistance towards the antimicrobials amongst the isolates of *K. pneumoniae* is an important health issue. The emergence of newly emerged *K. pneumoniae* strains shows resistance to drugs of choice. Therefore, novel strategies are required to prevent the transmission and contamination of *K. pneumoniae* strains.

Rapid diagnosis and treatment are necessary to manage resistant bacteria such as carbapenemase producing *K. pneumoniae* (Cp-Kp). For the control of Cp-Kp, the triple-combination remedy is being regarded as the best option for treatment. In such treatment, the polymyxin based remedy is the mainstay antibiotic.

The emergence of resistant bacterial strains has extremely constrained our capability for treating the diseases and therefore the development of new antibiotics is presently desirable. Traditional medicinal plants are credible to offer additional novel antibiotics in impending health concerns. The usage of plant extracts or pure natural substances in combination with conservative drugs might grip better promise for immediate and reasonable options for treatment. Certainly, some antibiotic remedies are now clinically accessible. For controlling the *K. pneumoniae* infections, the alternate strategies are employed like, vaccinations (capsular-polysaccharides, lipo-polysaccharides), traditional medications, bacteriophage remedy, plant derivative therapies of antibiotics, combinational chemotherapies of antimicrobial and plant extracts/compounds synergistic blends with conservative antibiotics.

Keywords: *Klebsiella pneumoniae, Enterobacteriaceae,* Antibiotics, Capsular-polysaccharides

1. INTRODUCTION

Klebsiella pneumoniae is a member of *Enterobacteriaceae*, lactose-fermenting, gram-negative bacilli, encapsulated, facultative anaerobic, nonmotile bacterium and is a member of the usual microbiota of numerous humans (Conlan, Kong, and Segre 2012). *K. pneumoniae* is a hospital acquired microbe which produces nosocomial pneumonia, urinary tract infection (UTI), intra-abdominal contaminations, pyogenic liver abscess (PLA), meningitis, bloodstream infection (BSI), ulcerative colitis, ankylosing spondylitis, small intestinal bowel overgrowth (SIBO), Crohn's disease, surgical and wound infections(Harris, Paterson, and Rogers 2015).

K. pneumoniae is a comparatively uncommon reason for community-acquired infections. In general, species of *Klebsiella* (*K. oxytoca* and *K. pneumoniae*) were the 3rdmost common microbes noticed amongst entire cases of central line-associated bloodstream infection (CLABSI), ventilator-associated pneumonia and catheter-associated urinary tract infection (CAUTI). *Klebsiella* species are widespread and colonize mammal's mucosal surfaces. In humans the colonization rates of *K. pneumoniae* vary ranging in the colon from 5-35%, oropharynx 1-5% and transiently is colonized the skin (Ko et al., 2002). In the Western nations, the liver abscess because of *K. pneumoniae* has correspondingly been increased (Fang et al., 2005). Invasive infections triggered through *K. pneumoniae* are related to other diseases like diabetes, cancer and earlier transplantation of organ (Meatherall et al., 2009).

In an investigation from the 1920-1960s, *K. pneumoniae* was a noteworthy reason for pneumonia (Carpenter 1990), though in the previous decade *K. pneumoniae* reasoned for <1% of pneumonia cases within North-America (Fang, Sandler, and Libby 2005) and about 10-50 cases of *K. pneumonia* were noticed yearly within the hospitals of US (Carpenter 1990). In medicinal textbooks, *K. pneumoniae* is listed as a significant reason for community-acquired pneumonia (Hughes, Sutherland, and Jones 1998). *K. pneumoniae* causes pneumonia in which there is an exacerbated inflammatory response, coupled with excessive infiltration of neutrophil and macrophage, excessive pro-inflammatory cytokines production and also lung get severely injured (Soares et al., 2006). A novel hyper-virulent clinical *K. pneumoniae* (hvKP) variant has appeared over the previous era (McCabe, Lambert, and Frazee 2010). *K. pneumoniae* is important healthcare related multidrug resistant microbe within the north-eastern US (Bratu et al., 2005). The most mutual mechanism of resistance is the b-*lactamase* production which hydrolyzes the carbapenem and is recognized as carbapenemase-*resistant K. pneumonia* (CKP) (Bradford et al., 2004). Carbapenem-resistant *K. pneumoniae* colonization has been related to numerous healthcare linked aspects (Schwaber et al., 2008). High mortality rates were noticed in patients affected by *K. pneumoniae* strain which shows resistance to carbapenem as compared to those affected by *K. pneumoniae*, which is sensitive to carbapenem (Schwaber et al., 2008).

To control infections that are triggered by carbapenem resistant strains of *K. pneumoniae*, optimum strategy within healthcare surroundings is unidentified. Initially when *K. pneumoniae* resistance to carbapenems was noticed, therefore numerous interventions for preventing transmission were presented comprising staff education, strengthening of hand cleanness, cleaning procedures, application of contact instructions for infected and colonized patients, usage of investigation cultures for identifying populated intensive care unit (ICU) patients. Novel drugs for controlling infections of multi drug resistant (MDR) *K. pneumoniae* are meropenem, vaborbactam, cefiderocol, aztreonam, avibactam, Imipenem, relebactam, ceftazidime, ceftaroline, plazomicin, eravacycline, cefepime and nacubactam (M. Bassetti et al., 2018). The progress of novel prophylactic and therapeutic approaches for controlling infections of *K. pneumoniae* is required. A novel isolated lytic phage-NK-5 was assessed against *K. pneumoniae* infection (Hung et al., 2011). Another strategy is the juice of cranberry used for preventing or eradicating colonization of *K. pneumoniae* within gut of hospitalized patients. From all natural dietary sources, the flavonoids were isolated and their activity was examined in combination with drugs as an approach against extended-spectrum beta-lactamase producing isolates of *K. pneumoniae* (Lin, Chin, and Lee 2005). In this book chapter, we have discussed several current management strategies for the control of infections triggered by *K. pneumoniae*.

2. STRATEGIES FOR CONTROL OF *KLEBSIELLA* ASSOCIATED INFECTIONS

2.1. Medication for *K. pneumoniae* Infections

K. pneumoniae is the main nosocomial microbe and produces extended-spectrum-beta lactamase (ESBL). Microbes which produce ESBL, frequently show resistance to carbapenem and several other antibiotics (Paterson and Bonomo 2005; Burns, Abadi, and Pirofski 2005). This bacterium hydrolyzed several antibiotics like cephalosporins, penicillins, carbapenems, monobactams, and β-lactamase inhibitors (Papp-Wallace et al., 2010). The co-resistance may be developed by the ESBL producers to further categories of anti-microbial substances like cotrimoxazole, aminoglycosides and fluoroquinolones. However, a recent study has reported the inhibition of *K. pneumoniae* via Fosfomycin (drug applied in treating UTI associated infections). Therefore, there new efficient antibiotics must be introduced which having low side effects, for instance, antibiotic's derived from natural plants (Liu et al., 2016).

2.1.1. Plants Suppressing K. pneumoniae Survival

Plant based medicine scheme performs an important part within the health care units. Approximately 80% of the population of the world depends chiefly on old medications for their healthcare (Owolabi, Omogbai, and Obasuyi 2007). As a source for novel medications, the potential of higher-plants is yet un-explored. Amongst the higher plants about 2,50,000-5,00,000 species have been assessed phyto-chemically. An extensive range of secondary metabolites is possessed by plants that shown some antimicrobial properties such as flavonoids, alkaloids, terpenoids, tannins, etc. (Dahanukar et al., 2000) (Table 1).

A larger quantity of antimicrobial agents is reported from medicinal plants. (Vashist and Jindal 2012). The clinical microbiologists are interested in the anti-microbial extracts of plants for two reasons i.e., firstly it is very probable that such phytochemicals search their way into the arsenal of antimicrobial agents prescribed by the physicians and numerous are also being checked on the humans, while secondly, the community is getting aware of issues associated with the mismanagement and over-prescription of antibiotics (Siani et al., 1999).

2.1.2. Plant Essential Oil Inhibiting K. pneumoniae

The plant extracts and essential oils (EOs) have been employed within aromatherapy and food-preservation (Buttner et al., 1996), perfume industries (Siani et al., 1999), and natural medicine remedies for many years (Lis-Balchin and Deans 1997). They are mainly derived chemically from oxygenated compounds and terpenes. Antiviral, antifungal, antioxidant, insecticidal and antibacterial properties are shown by the essential oils (Kordali et al., 2005). Inhibitory effects are shown by the EOs extracts from numerous plants against several types of microbes including fungi, bacteria and viruses. Numerous plant extracts have organic chemicals that show inhibitory action against particular microbes (Horne et al., 2001). Volatile oils or essential oils are greasy liquids that are taken from plants related materials like bark, twigs, leaves, herbs, fruits, seeds, flowers, buds, roots and wood (Table 2). By means of expression, extraction or fermentation they can be extracted but the steam distillation technique is more frequently applied for production at commercial level.

Biologically active substances are found in EOs (Hee Lee, Cho, and Lehrer 1997). One of the significant characteristics of EOs and components of essential oil is their hydrophobicity, which makes them able to separate the lipids of the cell membrane of bacteria and mitochondria, disturbing the cellular structures and making them more permeable which results in the leakage of molecules and ions leading bacterial cell death (Knobloch et al., 1986). The investigation has shown that the fourteen EOs producing plants are capable to prevent the growth of *K. pnuemoniae*. The thyme EO has shown potential anti-microbial action against several microbes. Results got by measuring minimum inhibitory concentration (MIC) indicated that essential oils from watered thyme-plants had MIC of 24.81, 11.34 and 8.12 mg/mL against *Escherichia coli*, *K. pneumoniae* and *K. oxytoca*, respectively.

Furthermore, *Klebsiella spp.* have shown susceptibility to even minor concentrations of essential oils (0.063 mg/L), which indicate the potential anti-bacterial activity of the thyme oil (Burt and Reinders 2003). Essential oils which were obtained from oregano, displayed anti-bacterial activity that was much common against *K. oxytoca* (with MICs of 2.11 and 0.90 mg/mL). Numerous species of *Klebsiella* have shown susceptibility to oregano essential oils and have about 0.5 mg/mL of MIC (Sahin et al., 2004).

2.1.3. Control of K. Pneumoniae via Withania Somnifera Leaf Extracts

Withania somnifera is considered as evergreen plant which has anti-oxidant (Bhattacharya, Satyan, and Ghosal 1997), pain-relieving (Kulkarni and Ninan 1997), anti-inflammatory (Hindawi et al.,1992), and memory enhancing effects (Schliebs et al., 1997). Several strains of *K. pneumonia* were found resistant to 3 antimicrobial substances i.e., erythromycin (66.6%), cefixime (60%) and ceftazidime (60%). In alternative study, *K. pneumonia* showed resistance to 6 antimicrobial substances including trimethoprim-sulfamethoxazole (25%), gentamicin (30%), nitrofurantoin (15%), nalidixicacid (15%), ampicillin (65%), and ciprofloxacin (20%). Zamani et al., (2013) stated that most of the antibiotics actively used against several spp of *K. pneumonia* are were ciprofloxacin (76.19%), amikacin (74.29%), ceftazidime (79.05%), tobramycin (79.05%), ceftriaxone (76.24%) and ceftizoxime (78.09%).

Subsequently numerous flavonoids including "quercetin glycosides" were likewise present inside the *W. somnifera* leaves (Kandil et al., 1994). It is supposed that the phenolic substances or flavonoids have selective inhibitory action. It is shown by in-vitro studies that the plant extracts prevented the growth of bacterial species, though with various efficiency. Ethanol which was extracted from *W. somnifera* has shown high inhibition of *K. pneumoniae* at higher concentrations.

2.2. Inhibition of *K. pneumoniae* using Iron Antagonizing Molecule and a Bacteriophage

K. pneumoniae is an important organism for biofilm-formation (Podschun and Ullmann 1998). Exo-polysaccharide matrix of the biofilm is degraded by depolymerases of phage (Hughes, Sutherland, and Jones 1998). Iron is the vital factor for bacterial growth which participates in oxygen and electron-transport mechanisms that is important for biofilm formation (Banin, Brady, and Greenberg 2006). Consequently, reducing the accessibility of iron is a possible means for damaging the development of biofilm by *K. pneumonia E. coli* and *Pseudomonas aeruginosa* (Hancock, Dahl, and Klemm 2010). Iron-antagonizing molecule has been used in combination with bacteriophage and separately for inhibiting biofilm formation of *K. pneumoniae* B5055 (Table 3). (Hancock, Dahl, and Klemm 2010) stated that Zn (II) and Co (II) possess a high iron affinity for the master controller protein of iron uptake leading to a reduction in the formation of biofilm through *E. coli*. In such an investigation, a considerable decrease in biofilm (1-3 days earlier) was noticed when 10 μM $FeCl_3$ and 500 μM $CoSO_4$ supplemented media was applied. This may be due to the increased metals concentrations that interfere with iron regulation through the shut-down of "Fur-controlled iron uptake systems" like aerobactin, ferric dicitrate and enterobactin. It also has some downstream impacts on putative adhesion factors that participate in biofilm development thus it results in harmful impacts on the formation of biofilm (Braun 2003). Consequently, the efficiency of depolymerase producing bacteriophage (i.e., KPO1K2) was investigated in eliminating *K. pneumoniae B5055* biofilms which were grownup inside the media containing iron and CoSO4 (about 500 μM). Complete eradication of the younger biofilms (up to 2 days old) was noticed by a combination treatment which was most probably because of the breakdown of exopolysaccharide matrix that surrounds the structure of biofilm by the depolymerase of bacteriophage (Verma, Harjai, and Chhibber 2009). Such findings suggest that the initial $CoSO_4$ addition and afterward treatment with bacteriophage that produces depolymerase is much more effective in biofilm degradation. Additionally, when different aged biofilms were separately treated with NDP (i-e non-depolymerase producing phage) and $CoSO_4$, so lesser eradication in the number of bacteria was noticed as compared to those biofilms which treatment was done with $CoSO_4$ and depolymerase producing bacteriophage.

2.3. Management of Pulmonary Infection of *K. pneumonia* by Probiotic *Bifidobacterium longum* 51A

The infection of *K. pneumoniae* is documented as a chief hazard to health because of the increased resistance to antibiotics. Alternate approaches should be required to improve immune system response so that to remove the bacteria might be efficient against pulmonary infection of *K. pneumoniae*. Probiotics are also appeared to be potential candidates. The species of *Bifidobacterium* which are nonpathogenic have been recognized as an effective anti-inflammatory probiotic with beneficial effects (Souza et al., 2013). More lately, it was confirmed that when *B. longum 51A* was orally given so it is capable of modulating the inflammatory-response elsewhere the gut (Vieira et al., 2015). Findings have shown that *B. longum* 51A exerts some significant immunomodulatory impacts that regulate injury of lung and dissemination of bacteria, which might be an effective remedy in controlling infections of *K. pneumoniae*. The immune system recognizes the microbes via stimulating PRRs (i-e pattern recognition receptors) for example the toll-like receptors (TLRs) (Medzhitov 2007). TLRs are the frontline pathogen sensors and the stimulation via pathogen associated molecular patterns (PAMP) is vital for rising an inflammatory response (Medzhitov 2007). Mal is defined as a bridge adapter that is accountable for the particular recruitment of MyD88 upon TLR4 or TLR2 receptor stimulation (Vatanavicharn et al., 2009). Mal has an important function in producing a defense against *K. pneumoniae* infection (Jeyaseelan et al., 2006). It was noticed that when *B. longum 51A* was orally given so it raised the ROS basal levels and it also caused in-vivo killing of *K. pneumoniae* via the alveolar macrophages. Such an outcome has shown that *B. longum 51A* changes the regulation of discrete signal via their recognition into alterations in alveolar macrophages which contribute for antibacterial action inside the lungs. Although it is not clear that by which mechanism such a species may directly contribute to signal within alveolar macrophages (Molloy, Bouladoux, and Belkaid 2012). Such data show that during infection of *K. pneumoniae*, when *B. longum 51A* is orally given so it helps to avoids inflammation of lungs, via stimulating bacterial innate immune sensing and hence facilitates the clearance of bacteria. Probiotic therapy of *B. longum 51A* might be an effective approach to improve the injury of the lung following infection of *K. pneumoniae* (Molloy, Bouladoux, and Belkaid 2012).

2.4. Topical Treatment of *K. pneumoniae* B5055 Induced Burn Wound Infection within Mice by means of Natural Products

Topical anti-microbial substances are crucial adjuncts used in treating and preventing burn wounds (Noronha C and Almeida A 2000). The extensive usage of these substances within the hospital has led to the appearance of MDR microbes (*Klebsiella*) producing severe types of opportunistic infections (Shukla et al., 2004). Currently, anti-microbial substances are not operative to treat such infections and unable to treat numerous bacterial infections because of the emergence of super resistant strains. Due to such reason, there is the exploration for novel anti-microbial substances, either by synthesis and design of novel agents or by natural sources investigation. Natural antimicrobial substances like bacteriophages, aloe vera and honey show effectiveness in topically treating burn wound infection.

2.4.1. Use of Honey against K. pneumoniae B5055

The antimicrobial substances are used to decrease a load of microbes within wounds and henceforth provide protection from infection. In the current investigation model of mice used having a burn wound so that to check the healing capability of honey and alovera gel in contrast to bacteriophage Kpn5 in infection treatment triggered via *K. pneumoniae* B5055. Honey is known to be the olden type of wound dressing (Molan and Betts 2004). Numerous in-vitro investigations showed that honey has an antibacterial action, and such observation is supported through clinical case investigations in which the honey was applied to harshly infected cutaneous wounds so it not just help out in the clearance of infection from the wound site but it also helped in healing of wound. Thus, honey is used against different types of bacteria comprising those which show extreme antibiotics resistance (McFadden et al., 2003). In such investigation, honey was used as an antimicrobial substance for treating burn wound infection within mice triggered via the *K. pneumoniae* B5055. The sterile honey antibacterial action was noticed to be heat stable and it stayed active even afterward autoclaving process. In vitro the undiluted honey displayed the antibacterial action against *K. pneumoniae* B5055. Regarding earlier investigations, Manuka honey showed bacteriogenic action against diverse type of MDR microbes including *K. pneumoniae* which produces extended spectrum b-lactamases (Shah Pratibha and Williamson Manita 2015).

2.4.2. Alovera Gel

Amongst therapeutic plants, aloe vera or *aloe barbadensis* is used for therapeutic purposes for centuries. Certain medicinal products and cosmetics are made-up from the viscous tissue of AVG (aloe vera gel) situated in the middle of a leaf of the aloe vera (Reynolds and Dweck 1999). Aloe vera gel possesses wound healing action and polysaccharides of aloe have beneficial influences in the prevention of burns. It is both anti-inflammatory and antimicrobial (Schulz et al., 2004). The aloe vera gel antimicrobial action was checked in such a study for treating burn wound infection of mice triggered through *K. pneumoniae B5055*. The unsterilized aloe vera gel antibacterial action was noticed in-vitro against *K. pneumoniae B5055* and once it was topically used at the burned spot so considerably better percent survival of 26.66% was noticed in mice which received the treatment as compared to untreated mice on the 7th day after treatment. Furthermore, the aloe vera extract antibacterial constituents were operative against *E. coli, K. pneumoniae, Streptococcus pyogenes, Propionibacterium acne, S. aureus, Salmonella typhi, P. aeruginosa* and *Helicobacter pylori* (Pugh et al., 2001).

2.5. New Management and Treatment of *Klebsiella pneumonia*

2.5.1. Combination Therapy

Several investigations have predominantly described the combination remedy of polymyxin-carbapenem in *K. pneumonia* patients (MIC ≤ 8–16) (Tumbarello et al., 2015). Some past studies evaluated individuals who were affected with Cr-Kp having higher degree resistance to carbapenem (Satlin et al., 2017). MIC based variations in clinical outcomes were supported through in-vitro testing showing synergism amongst the bactericidal action of meropenem plus polymyxin B against the isolates of *K. pneumoniae* (Diep et al., 2017). When bacterial isolates were exposed to such drugs combination resulting in morphological alterations within bacterial isolates. Hence, polymyxins combinations along with the optimized dosage of a carbapenem were shown to be effective for treating infections produced by Cr-Kp strains (Pournaras et al., 2011). Alternative investigation recommends that tigecycline, doxycycline and rifampicin might be beneficial additions to polymyxin-B in treating infections triggered by *K. pneumoniae* (Endimiani et al., 2010). *In-vitro* synergistic action was shown by the rifampicin and Polymyxin B against 15 isolates of *K. pneumoniae* which were carbapenem resistant (Bratu et al., 2005) (Table 4).

2.5.1.1. Double Carbapenem Approach for the Treatment and Management of Severe Carbapenemase-producing *K. pneumoniae* Infections

Over previous years, many resistance mechanisms were shown by *K. pneumoniae*. Afterward its initial isolation in the U.S in 1996, the carbapenem resistant *K. pneumoniae* (Cr-Kp) strains triggered outbreaks of nosocomial infections within Greece, Italy, Israel and northeastern United (Matteo Bassetti et al., 2015). The Cr-Kp strains have been spread worldwide and posing a huge threat to the health of people. Infections associated with *K. pneumonia* are related to higher mortality and increased cost to cure such infections (Gutierrez-Gutierrez et al., 2016). There are drug resistant phenotypes of Cr-Kp and the usage of some rescue drugs like aminoglycosides, fosfomycin, tigecycline, and colistin is are often linked with clinical disappointment. Though several investigations have shown that in-vitro when more than one active drug is used in combination so it might be better than mono-therapy (Jarlier et al., 2019). On the other hand, some novel antimicrobials such as beta-lactamase/ beta-lactam inhibitor combinations i.e., ceftazidime-avibactam is yet poor and it is linked with resistance emergence in the treatment process (Zusman et al., 2017). Currently, synergistic combination of 2 carbapenems i.e., ertapenem in combination with either imipenem, doripenem or meropenem, separately or in combination with some other drugs was suggested as a treatment for Cr-

Kp. Individuals who were seriously infected with Cr-Kp when treated with a double carbapenem remedy comprising ertapenem showed decrease mortality. Besides reduced mortality, eradication of microbes was also observed in individuals who were treated with double carbapenem therapy (Cprek and Gallagher 2015).

Carbapenemases *producing K. pneumoniae* generally show resistance to the entire beta lactams and other drugs such as fluoroquinolones, aminoglycosides, and sulphonamides (Matteo Bassetti et al., 2015). However, the usage of colistin and gentamicin is often limited because it causes renal toxicity (Shields et al., 2016). The double carbapenem (DC) regimen is useful in a way like; ertapenem binds with a higher affinity to the active site of carbapenemase of *K. pneumoniae* and capable to avoid the breakdown of the given antibiotic i.e., doripenem, meropenem or imipenem that might preserve bactericidal action against the infectious agent. The initial observation of DC effectiveness was confirmed via Bulik and Nicolau in 2011 comprising ertapenem plus doripenem or meropenem within patients treated for an infection of CR-Kp (Bulik and Nicolau 2011).

2.5.1.2. Triple Combination Antibiotic Therapy for Carbapenemases-Producing *K. pneumoniae*

Amongst *Enterobacteriaceae*, carbapenemase production is common in *K. pneumoniae* (Falagas et al., 2014). Recently there exists no suitable treatment for *K. pneumoniae* infections ((Falagas et al., 2014) A current review has shown that the usage of combination therapy for Cp-Kp is encouraging (Tzouvelekis et al., 2014). When the triple drug combination therapy was invitro analyzed so it showed positive results for the treatment of Cr-Kp (Diep et al., 2017). It is shown that the triple drug combination remedy for treatment of Cp-Kp is applied with polymyxin. In- vitro combinations of drugs prevented the resistance development and hence it inspired the clinicians to discover such promising combinations of drugs for the treatment of patients (Diep et al., 2017). Commonly applied combinations for treatment were tigecycline combined with a carbapenem or colistin-polymyxin B. Usage of combination remedy outcome resulted in low mortality of about 12.5% as compared to about 66.7% with a carbapenem, polymyxin, or tigecycline separately (Qureshi et al., 2012). Usage of combinations remedy provided better survival in bacteremia which was caused by *K. pneumoniae*. The colistin-polymyxin B was a backbone drug in triple drug combination remedy along with tigecycline indicating that polymyxins are too much important for treating the infections caused by highly drug resistant microbes (Arutselvi et al., 2012). Findings have shown that polymyxins were applied as a component of triple drug combination remedy in the treatment of Cp-Kp. Numerous studies have stated the emergence of resistance to polymyxins with increased mortality (Ahmed, Raman, and Veerappan 2016). The development of resistance to polymyxin is alarming. Resistance to polymyxins might be produced by the alterations in the LPS of an outer membrane or when the capsular polysaccharides are highly produced (Velkov et al., 2010). During in-vitro analysis, ceftazidime-avibactam showed potential action against the carbapenem-resistant Enterobacteriaceae (CRE) isolates (Sader et al., 2015). Further combinations of β-lactamase/ β-lactam inhibitor are investigated counting the aztreonam–avibactam and ceftolozane–tazobactam (Papp-Wallace and Bonomo 2016). Plazomicin (which is innovative aminoglycoside) has shown activity against CRE and it is in Phase-3 clinical trial as a component of combination remedy (Galani et al., 2012).

2.6. Other Antibiotics for the Treatment and Management of *Klebsiella pneumonia*

Following other antibiotics are used for treatment and control of infections caused by the *K. pneumonia* (Table 4).

2.6.1. Polymyxins
In-vitro sensitivity to polymyxins (i.e., polymyxin B and colistin) among isolates of carbapenemase producing *enterobacteriaceae* (CPE) varies worldwide from 80 to 100%. Although, within certain regions, resistance could be higher because of resistant strains clonal spread (Bogdanovich et al., 2011). As compared to polymyxin B, the colistin

commonly used (Dudhani et al., 2010). Colistin is the only agent that is effective against CPE and it can help in treating bloodstream infections (Arnold et al., 2011). Colistin resistance may arise most commonly within Cr-Kp as compared to MDR *P. aeruginosa* or *A. baumannii* (Samonis et al., 2010). Frequent usage of such agents is linked with the development of hetero-resistant isolates because of the modification of the LPS membrane (Shrivastava, Varma, and Padh 2011; Poudyal et al., 2008).

2.6.2. Aminoglycosides

Resistance to aminoglycoside is rising amongst the isolates of CPE. Within susceptible strains, in-vitro data showed rapid bactericidal action of gentamicin against susceptible strains (Bratu et al., 2005). Other lineages might carry gentamicin modifying enzymes and other aminoglycosides, for example, tobramycin and amikacin, which have been revealed to be less efficient against infections caused by MDR *K. pneumoniae*. Published data about the aminoglycosides usage as monotherapy against Cp-Kp infections are uncommon and hence could not be suggested.

2.6.3. Fosfomycin

It is a natural derivative of phosphonic acid which inhibits the cell wall biosynthesis at the very initial stage as compared to beta-lactam drugs. Invitro this antibiotic has activity against extended spectrum beta-lactamase (ESBL) producing *Enterobacteriaceae* counting Cr-Kp (Endimiani et al., 2010). The fosfomycin range of activity was checked against 68 isolates of *K. pneumoniae* which were KPC producers. Among these, 23 shown non susceptibility to colistin or tigecycline. The rates of susceptibility were 93% for the entire group and it was 87% for the group which was non-susceptible to colistin or tigecycline and 83% for the high drug-resistant group i.e., non-susceptible to colistin and tigecycline (Endimiani et al., 2010) (Kurdekar, Hegde, and Hebbar 2012) When the fosfomycin is applied in combination with some another drug such as gentamicin, colistin, and tazobactam, then it is effective in treating the *K. pneumoniae* (Falcone and Paterson 2016).

2.7. New Antimicrobial Agents

In 2015, the US Food and Drug Administration (FDA) approved the ceftazidime-avibactam for treating UTI and pyelonephritis and can be applied along with metronidazole for treating serious intraabdominal infections. Avibactam is a novel non-beta-lactam drug that has shown effectiveness against class C and A beta-lactamases such as KPC, AmpC and ESBLs (Castanheira, Mendes, and Sader 2017). The ceftazidime-avibactam possess action against isolates of MDR *K. pneumonia*. No data is available on the usage of ceftazidime-avibactam for treating Cr-Kp infections, but available data show that ceftazidime-avibactam might be a good therapeutic choice for several patients having Cr-Kp infection (Trabattoni et al., 2017).

2.7.1. Management by Novel Agents

Certain novel b-lactamase inhibitors which can be not hydrolyzed by class-A carbapenemases and ESBLs are presently in the developmental stage. Such novel drugs comprise BLI-489, LK-157, and NXL-104. NXL-104 has shown activity against 6 diverse KPC-producing isolates (Stachyra et al., 2009). LK-157 is a new tricyclic carbapenem that possesses effectiveness against class C and A b-lactamases (Paukner et al., 2009). BLI-489 is a bicyclic penem molecule having activity against an extensive range of enzymes but its activity is not assessed precisely against KPC-producing microbes (Pradhan, Panda, and Tripathy 2009). ACHN-409 is a novel generation of aminoglycoside i.e., neoglycoside and in-vitro it has shown effectiveness against KPC-producing strains (Verma, Harjai, and Chhibber 2009).

2.7.1.1. By Using Gentamicin Nanoparticles, The Clearance of Intracellular *K. pneumoniae* Infection

K. pneumoniae may survive within the cell even afterward phagocytosis through limiting lysosomes fusion with the *Klebsiella* containing vacuole (KCV) and creates an intracellular infection reservoir (Cano et al., 2015). To clinically treat such intracellular infection is a challenging task; for e.g., aminoglycosides, which could treat successfully the extracellular infections of *K. pneumoniae*, but its poor penetration into cells and insufficient subcellular distribution, have reduced its activity towards intracellular infection (Abraham and Walubo 2005). Novel drugs and new drug delivery approaches are required to treat lethal and persistent pathogens. The ability to improve the drug delivery into *K. pneumonia* infected macrophages was checked using a nanoformulation strategy (Bains et al., 2016). The nanoparticle delivery system of the drug for treating *K. pneumoniae* infection has been examined formerly. Nanoparticle delivery systems for *K. pneumoniae* treatment were investigated. These comprise the use of metal nanoparticles like silver or gold which possess antimicrobial properties (Zazo, Colino, and Lanao 2016). Drugs loaded nanoparticle systems are also investigated comprising gentamicin loaded fucoidan/chitosan nanoparticles or ceftazidime-loaded liposomes to improve drugs pulmonary delivery (Y. C. Huang et al., 2016).

2.7.1.2. Gentamicin-Loaded NPs in Management of *K. pneumoniae* Infections

The model of *Galleria mellonella* larvae was in-vivo applied so that to evaluate the antimicrobial action of gentamicin loaded nanoparticles (GNPs) in treating infection of *K. pneumoniae*. The model of *G. mellonella* is commonly applied for pharmacokinetic investigations and antibiotic susceptibility (Hill, Veli, and Coote 2014). Subsequently, in a study, larvae of *G. mellonella* infected with (1×10^4) *Klebsiella*, and afterward free solution of gentamicin (3.6 μg/larvae), Gentamicin loaded NPs (40 μg/larvae) at equivalent drug concentration), and also PBS (10 μl) or non-drug loaded NPs were given. Then the larvae survival was checked and it was found that in the existence of non-drug loaded NPs or PBS the *G. mellonella* larvae succumbed to the infection (by 96 h), while on the other hand when it was treated with GNPs or free drug then the survival was suggestively improved. Results obtained from such a study emphasized that the formulation of nanoparticle was operative just like the free drug in this in-vivo infection model (Lu et al., 2018). GNPs might protect against the infection of *K. pneumoniae* in *G. larvae*. Larvae of *G. mellonella* were prophylactically treated with either GNPs or free drugs up to 5 days former to bacterial challenge and then the survival was checked for 5 days. This trial has shown that while a 24 hour initial treatment of both GNPs and free drug produced no considerable variations in survival to infection but it got obvious that with both 48 hour and 72 hour initial treatments, GNPs offered improved protection (Schentag and Jusko 1977).

2.8. Combinations of Extracts with Antimicrobials in Management of *K. pneumoniae* Infections

Antimicrobial agents operative for treating infection of MDR bacteria are generally limited. Hence, there is a need to search substances that potentiate the anti-microbial action of drugs on such bacterial specie. Plant extracts in synergism with antibiotics are considered as a novel strategy that helps to solve the issue of resistance in bacteria (Aiyegoro and Okoh 2009). Phytotherapy has numerous benefits linked with the synergistic associations such as enhanced efficiency, reduced undesirable impacts, enhanced stability, and therapeutic effect even with comparatively smaller dosages (Aiyegoro and Okoh 2009).

2.8.1. Synergistic Consequence of Thymbra spicata L. Extracts with Antibiotics in Management of K. pneumoniae Strains

T. spicata, (Lamiaceae), is a native plant of Syria (Mouterde et al., 1983). It is applied in medications and it is a medicinal plant. The essential oils of *T. spicata* has both antioxidant and antibacterial properties. This plant is used to treat sore throat and respiratory infection. It gives a better taste and flavors to different types of meals and is also used as a spice (Meletis et al., 2011). The antimicrobial actions of this plant extract were tested against different bacteria (Celiktas et al., 2007). *K. pneumoniae* causes nosocomial and community acquired infections for example meningitis, pyogenic liver abscess, and pneumonia (Porsche and Ullmann 1998). Antimicrobial drugs are limited to treat the MDR bacterial infections, hence plant extracts in combination with antimicrobials are considered as a new approach

(Khameneh et al., 2016). In a study, the plant extracts MICs against the strains of *K. pneumoniae* and MDR *S. aureus* varied from 12.5 to 100 and 6.25 to 50 mg/ml (Nascimento et al., 2000). Omar et al., (20130 described that *K. pneumoniae* and *S. aureus* showed sensitivity to aqueous extract of *T. spicata* at MIC of 50 mg/ml but *K. pneumoniae* showed resistance to ethanolic extract. The current investigation was focused on the synergistic and antibacterial action of plant extracts with drugs. The results indicated the synergistic and additive associations between phytochemicals of *T. spicata* and antimicrobials, which are protein synthesis inhibitor (amikacin), DNA synthesis inhibitor (ciprofloxacin), and cell wall inhibitors (ampicillin, cefotaxime). Beta-lactam drugs were only agents which were linked with good synergistic potential and therefore were approved in such investigation (Hübsch et al., 2014). The synergistic associations observed between the antibiotics and extracts of *T. spicata* could be translated into beneficial uses in *K. pneumonia* and *S. aureus* infections (Barakat et al., 2013). The synergistic activity of *T. spicata* extracts might be due to the active compound such as carvacrol which causes disturbance within the cell wall and causes depolarization of the cytoplasmic membrane which enables the entry of the drug into the bacterial cell (J. Xu et al., 2008).

2.9. Photodynamic Therapy (PDT) in Management of *K. pneumoniae* Infections

Photodynamic therapy (PDT) is used for treating the infections produced by *ESBL* producing *K. pneumoniae*. PDT involves visible light of proper wavelength on photosensitizers (PS). PS is then converted into an excited state, and later on, the reactive oxygen species (ROS) are formed called singlet oxygen. Bacterial killing is done in un-specific manner by the ROS and thus the resistance to PDT would be not likely to develop (Wood et al., 2006). As a 1st choice of drug for treating Cr-Kp higher carbapenem doses are used, or the same is used in combination with a variety of other drugs like tigecycline, colistin, fosfomycin or gentamicin. In-vitro analysis of several new generations of cephalosporins has shown activity against Cr-

Kp, when used in combination with tazobactam. Although other approaches such as antimicrobial photodynamic inactivation (aPDI) should be suggested along with the advent of novel drugs. Photodynamic therapy is discovered against the ESKAPE pathogens (ESKAPE pathogens: *E. faecium, S. aureus, K. pneumoniae, A. baumannii, P. aeruginosa,* and *Enterobacter* species). Numerous kinds of light-activated compounds are studied in photoinactivation of *K. pneumoniae* like phthalocyanines, porphyrin derivatives and their precursors such as (5-ALA, MAL), boron dipyrromethene (BODIPY)-based, phenothiazines, and phenothiazines. Currently, indole derivatives were studied which represent a novel structure scaffold for application of aPDI. These compounds have shown activity against gram-negative species only when polymyxin E was present (Edwards et al., 2018) The photophrin is FDA approved drug for the treatment of cancers was also found effective against *K. pneumonia* when potassium iodide (KI) was present (approximately 6-log10 reduction was observed in viable cells of *K. pneumoniae*) (L. Huang et al., 2017). When the MAL or 5-ALA was applied more than 3-log10 reductions were observed against ESBL producing or non-producing strains (Liu et al., 2016). In PT when vitamin K5 was used as a PS, so the combination of ultraviolet A and vitamin K5 caused 7-logs reductions in the survival of *K. pnuemoniae* but it too resulted in a reduction of some other gram-negative species (F. Xu et al., 2018).

CONCLUSION

K. pneumonia is a huge threat to public health in hospital settings because of the emergence of MDR and the development of hyper-virulent strains. Due to the frequent emergence of high-risk isolates and the worldwide spread of such isolates have made it very difficult for clinicians in finding suitable therapeutic options. The management of KPC-KP infections within seriously infected individuals relies on concerted multidimensional therapy approaches. The proper therapeutic regimes for infections of KPC-producing isolates are unknown. Presently, very limited information is available from *in-vitro* infection models or animal, therefore there is a need of further research. Clinical outcomes and observational studies are instantly required so that to determine a wise and suitable therapeutic regime for KPC infections. Last of all, infections triggered by KPC-producing microorganisms further highlight the need to investigate combination remedy and rational treatment approaches. Innovative approaches like PDT and bacteriophage showed promising anti-biofilm and antimicrobial activities against *K. pneumoniae* which can be further investigated as therapeutic options in treating infections triggered by *K. pneumoniae*.

REFERENCES

Abraham, A. M., and A. Walubo. 2005. "The Effect of Surface Charge on the Disposition of Liposome-Encapsulated Gentamicin to the Rat Liver, Brain, Lungs and Kidneys after Intraperitoneal Administration." *International Journal of Antimicrobial Agents* 25 (5). https://doi.org/10.1016/j.ijantimicag.2005.01.018.

Ahmed, Khan Behlol Ayaz, Thiagarajan Raman, and Anbazhagan Veerappan. 2016. "Future Prospects of Antibacterial Metal Nanoparticles as Enzyme Inhibitor." *Materials Science and Engineering C*. https://doi.org/10.1016/j.msec.2016.06.034.

Aiyegoro, O. A., and A. I. Okoh. 2009. "Use of Bioactive Plant Products in Combination with Standard Antibiotics: Implications in Antimicrobial Chemotherapy." *Journal of Medicinal Plants Research*.

Arutselvi, R., T. Balasaravanan, P. Ponmurugan, P. Suresh, and N. Ramachandran. 2012. "Comparative Studies of Anti-Microbial Activity of Turmeric and Selected Medicinal Plant Leaf Extracts Used in Indian Traditional Medicine." *Journal of Herbs, Spices and Medicinal Plants* 18 (3). https://doi.org/10.1080/10496475.2012.680245.

Banin, Ehud, Keith M. Brady, and E. Peter Greenberg. 2006. "Chelator-Induced Dispersal and Killing of Pseudomonas Aeruginosa Cells in a

Biofilm." *Applied and Environmental Microbiology* 72 (3). https://doi.org/10.1128/AEM.72.3.2064-2069.2006.

Barakat, Ahmad, Lara Hanna Wakim, Nelly Arnold Apostolides, Ghassan Srour, and Marc el Beyrouthy. 2013. "Variation in the Essential Oils of Thymbra Spicata L. Growing Wild in Lebanon According to the Date of Harvest." *Journal of Essential Oil Research* 25 (6). https://doi.org/10.1080/10412905.2013.809321.

Bassetti, M., D. R. Giacobbe, H. Giamarellou, C. Viscoli, G. L. Daikos, G. Dimopoulos, F. G. de Rosa, et al., 2018. "Management of KPC-Producing Klebsiella Pneumoniae Infections." *Clinical Microbiology and Infection*. https://doi.org/10.1016/j.cmi.2017.08.030.

Bassetti, Matteo, Jan J. de Waele, Philippe Eggimann, Josè Garnacho-Montero, Gunnar Kahlmeter, Francesco Menichetti, David P. Nicolau, et al., 2015. "Preventive and Therapeutic Strategies in Critically Ill Patients with Highly Resistant Bacteria." *Intensive Care Medicine*. https://doi.org/10.1007/s00134-015-3719-z.

Buttner, M. P., Willeke, K., Grinshpun, S. A. (1996): *Sampling and analysis of airborne microorganisms.* In: Hurst CJ, Knudsen GR, McInerney MJ, Stetzenbach LD, Walter.

Bhattacharya, Salil K., Kalkunte S. Satyan, and Shibnath Ghosal. 1997. "Antioxidant Activity of Glycowithanolides from Withania Somnifera." *Indian Journal of Experimental Biology* 35 (3).

Bogdanovich, Tatiana, Jennifer M. Adams-Haduch, Guo Bao Tian, Minh Hong Nguyen, Eun Jeong Kwak, Carlene A. Muto, and Yohei Doi. 2011. "Colistin-Resistant, Klebsiella Pneumoniae Carbapenemase (KPC)-Producing Klebsiella Pneumoniae Belonging to the International Epidemic Clone ST258." *Clinical Infectious Diseases* 53 (4). https://doi.org/10.1093/cid/cir401.

Bradford, P. A., S. Bratu, C. Urban, M. Visalli, N. Mariano, D. Landman, J. J. Rahal, S. Brooks, S. Cebular, and J. Quale. 2004. "Emergence of Carbapenem-Resistant Klebsiella Species Possessing the Class A Carbapenem-Hydrolyzing KPC-2 and Inhibitor-Resistant TEM-30 - Lactamases in New York City." *Clinical Infectious Diseases* 39 (1). https://doi.org/10.1086/421495.

Bratu, Simona, David Landman, Robin Haag, Rose Recco, Antonella Eramo, Maqsood Alam, and John Quale. 2005. "Rapid Spread of Carbapenem-Resistant Klebsiella Pneumoniae in New York City: A New Threat to Our Antibiotic Armamentarium." *Archives of Internal Medicine* 165 (12). https://doi.org/10.1001/archinte.165.12.1430.

Braun, Volkmar. 2003. "Iron Uptake by Escherichia Coli." *Frontiers in Bioscience*. https://doi.org/10.2741/1232.

Bulik, Catharine C., and David P. Nicolau. 2011. "Double-Carbapenem Therapy for Carbapenemase-Producing Klebsiella Pneumoniae." *Antimicrobial Agents and Chemotherapy* 55 (6). https://doi.org/10.1128/AAC.01420-10.

Burns, Tamika, Maria Abadi, and Liise Anne Pirofski. 2005. "Modulation of the Lung Inflammatory Response to Serotype 8 Pneumococcal Infection by a Human Immunoglobulin M Monoclonal Antibody to Serotype 8 Capsular Polysaccharide." *Infection and Immunity* 73 (8). https://doi.org/10.1128/IAI.73.8.4530-4538.2005.

Burt, S. A., and R. D. Reinders. 2003. "Antibacterial Activity of Selected Plant Essential Oils against Escherichia Coil O157:H7." *Letters in Applied Microbiology* 36 (3). https://doi.org/10.1046/j.1472-765X.2003.01285.x.

Buttner, M. P., Willeke, K., Grinshpun, S. A. (1996): *Sampling and analysis of airborne microorganisms*. In: Hurst CJ, Knudsen GR, McInerney MJ, Stetzenbach LD, Walter.

Cano, Victoria, Catalina March, Jose Luis Insua, Nacho Aguiló, Enrique Llobet, David Moranta, Verónica Regueiro, et al., 2015. "Klebsiella Pneumoniae Survives within Macrophages by Avoiding Delivery to Lysosomes." *Cellular Microbiology* 17 (11). https://doi.org/10.1111/cmi.12466.

Carpenter, John L. 1990. "Klebsiella Pulmonary Infections: Occurrence at One Medical Center and Review." *Reviews of Infectious Diseases* 12 (4). https://doi.org/10.1093/clinids/12.4.672.

Castanheira, Mariana, Rodrigo E. Mendes, and Helio S. Sader. 2017. "Low Frequency of Ceftazidime-Avibactam Resistance among Enterobacteriaceae Isolates Carrying BlaKPC Collected in U.S.

Hospitals from 2012 to 2015." *Antimicrobial Agents and Chemotherapy* 61 (3). https://doi.org/10.1128/AAC.02369-16.

Celiktas, O. Yesil, E. E. Hames Kocabas, E. Bedir, F. Vardar Sukan, T. Ozek, and K. H. C. Baser. 2007. "Antimicrobial Activities of Methanol Extracts and Essential Oils of Rosmarinus Officinalis, Depending on Location and Seasonal Variations." *Food Chemistry* 100 (2). https://doi.org/10.1016/j.foodchem.2005.10.011.

Conlan, Sean, Heidi H. Kong, and Julia A. Segre. 2012. "Species-Level Analysis of DNA Sequence Data from the NIH Human Microbiome Project." *PLoS ONE* 7 (10). https://doi.org/10.1371/journal.pone.0047075.

Chobba, I. B., Bekir A., Mansour, R. B., Drira, N., Gharsallah, N., Kadri, A. (2012): In vitro Evaluation of Antimicrobial and Cytotoxic Activities of Rosmarinus officinalis L. (Lamiaceae) Essential Oil Cultivated from SouthWest Tunisia. *Japs*. 2:034-039.

Cooper, C. J., Koonjan, S., & Nilsson, A. S. (2018): Enhancing whole phage therapy and their derived antimicrobial enzymes through complex formulation. *Pharmaceuticals* 11: 34.

Diep, John K., David M. Jacobs, Rajnikant Sharma, Jenna Covelli, Dana R. Bowers, Thomas A. Russo, and Gauri G. Rao. 2017. "Polymyxin B in Combination with Rifampin and Meropenem against Polymyxin B-Resistant KPC-Producing Klebsiella Pneumoniae." *Antimicrobial Agents and Chemotherapy* 61 (2). https://doi.org/10.1128/AAC.02121-16.

Edwards, Leslie, Danielle Turner, Cody Champion, Megha Khandelwal, Kailee Zingler, Cassidy Stone, Ruwini D. Rajapaksha, et al., 2018. "Photoactivated 2,3-Distyrylindoles Kill Multi-Drug Resistant Bacteria." *Bioorganic and Medicinal Chemistry Letters* 28 (10): 1879–86. https://doi.org/10.1016/j.bmcl.2018.04.001.

Endimiani, Andrea, Gopi Patel, Kristine M. Hujer, Mahesh Swaminathan, Federico Perez, Louis B. Rice, Michael R. Jacobs, and Robert A. Bonomo. 2010. "In Vitro Activity of Fosfomycin against BlaKPC-Containing Klebsiella Pneumoniae Isolates, Including Those Nonsusceptible to Tigecycline and/or Colistin." *Antimicrobial Agents* and Chemotherapy 54 (1): 526–29. https://doi.org/10.1128/AAC. 01235-09.

Falagas, Matthew E., Panagiota Lourida, Panagiotis Poulikakos, Petros I. Rafailidis, and Giannoula S. Tansarli. 2014. "Antibiotic Treatment of Infections Due to Carbapenem-Resistant Enterobacteriaceae: Systematic Evaluation of the Available Evidence." *Antimicrobial Agents and Chemotherapy* 58 (2): 654–63. https://doi.org/10.1128/AAC.01222-13.

Falcone, Marco, and David Paterson. 2016. "Spotlight on Ceftazidime/Avibactam: A New Option for MDR Gram-Negative Infections." *Journal of Antimicrobial Chemotherapy* 71 (10): 2713–22. https://doi.org/10.1093/jac/dkw239.

Fang, Ferric C., Netanya Sandler, and Stephen J. Libby. 2005. "Liver Abscess Caused by MagA+ Klebsiella Pneumoniae in North America." *Journal of Clinical Microbiology* 43 (2): 991–92. https://doi.org/10.1128/JCM.43.2.991-992.2005.

Galani, Irene, Maria Souli, George L. Daikos, Zoi Chrysouli, Garyphalia Poulakou, Mina Psichogiou, Theofano Panagea, et al., 2012. "Activity of Plazomicin (ACHN-490) against MDR Clinical Isolates of Klebsiella Pneumoniae, Escherichia Coli, and Enterobacter Spp. from Athens, Greece." *Journal of Chemotherapy* 24 (4): 191–94. https://doi.org/10.1179/1973947812Y.0000000015.

Hancock, Viktoria, Malin Dahl, and Per Klemm. 2010. "Abolition of Biofilm Formation in Urinary Tract Escherichia Coli and Klebsiella Isolates by Metal Interference through Competition for Fur." *Applied and Environmental Microbiology* 76 (12). https://doi.org/10.1128/AEM.00241-10.

Harris, Patrick, David Paterson, and Benjamin Rogers. 2015. "Facing the Challenge of Multidrug-Resistant Gram-Negative Bacilli in Australia." *Medical Journal of Australia*. https://doi.org/10.5694/mja14.01257.

Hee Lee, I. N., Yoon Cho, and Robert I. Lehrer. 1997. "Effects of PH and Salinity on the Antimicrobial Properties of Clavanins." *Infection and Immunity* 65 (7). https://doi.org/10.1128/iai.65.7.2898-2903.1997.

Hill, Lucy, Neyme Veli, and Peter J. Coote. 2014. "Evaluation of Galleria Mellonella Larvae for Measuring the Efficacy and Pharmacokinetics of Antibiotic Therapies against Pseudomonas Aeruginosa Infection." *International Journal of Antimicrobial Agents* 43 (3). https://doi.org/10.1016/j.ijantimicag.2013.11.001.

Horne, Diane, Mitchel Holm, Craig Ober, Sue Chao, and D. Gary Young. 2001. "Antimicrobial Effects of Essential Oils on Streptococcus Pneumoniae." *Journal of Essential Oil Research* 13 (5). https://doi.org/10.1080/10412905.2001.9712241.

Huang, Liyi, Grzegorz Szewczyk, Tadeusz Sarna, and Michael R. Hamblin. 2017. "Potassium Iodide Potentiates Broad-Spectrum Antimicrobial Photodynamic Inactivation Using Photofrin." *ACS Infectious Diseases* 3 (4). https://doi.org/10.1021/acsinfecdis.7b00004.

Huang, Yi Cheng, Rou Ying Li, Jiun Yu Chen, and Jen Kun Chen. 2016. "Biphasic Release of Gentamicin from Chitosan/Fucoidan Nanoparticles for Pulmonary Delivery." *Carbohydrate Polymers* 138. https://doi.org/10.1016/j.carbpol.2015.11.072.

Hübsch, Z., R. L. van Zyl, I. E. Cock, and S. F. van Vuuren. 2014. "Interactive Antimicrobial and Toxicity Profiles of Conventional Antimicrobials with Southern African Medicinal Plants." *South African Journal of Botany* 93. https://doi.org/10.1016/j.sajb.2014.04.005.

Hughes, Kevin A., Ian W. Sutherland, and Martin v. Jones. 1998. "Biofilm Susceptibility to Bacteriophage Attack: The Role of Phage-Borne Polysaccharide Depolymerase." *Microbiology* 144 (11). https://doi.org/10.1099/00221287-144-11-3039.

Hung, Chih Hsin, Chih Feng Kuo, Chiou Huey Wang, Ching Ming Wu, and Nina Tsao. 2011. "Experimental Phage Therapy in Treating Klebsiella Pneumoniae-Mediated Liver Abscesses and Bacteremia in Mice." *Antimicrobial Agents and Chemotherapy* 55 (4). https://doi.org/10.1128/AAC.01123-10.

Jarlier, Vincent, Liselotte Diaz Högberg, Ole E. Heuer, José Campos, Tim Eckmanns, Christian G. Giske, Hajo Grundmann, et al., 2019. "Strong Correlation between the Rates of Intrinsically Antibiotic-Resistant Species and the Rates of Acquired Resistance in Gram-Negative Species

Causing Bacteraemia, EU/EEA, 2016." *Euro Surveillance : Bulletin Europeen Sur Les Maladies Transmissibles = European Communicable Disease Bulletin* 24 (33). https://doi.org/10.2807/1560-7917.ES.2019.24.33.1800538.

Jeyaseelan, Samithamby, Scott K. Young, Masahiro Yamamoto, Patrick G. Arndt, Shizuo Akira, Jay K. Kolls, and G. Scott Worthen. 2006. "Toll/IL-1R Domain-Containing Adaptor Protein (TIRAP) Is a Critical Mediator of Antibacterial Defense in the Lung against Klebsiella Pneumoniae but Not Pseudomonas Aeruginosa." *The Journal of Immunology* 177 (1). https://doi.org/10.4049/jimmunol.177.1.538.

Kandil, F. E., N. H. el Sayed, A. M. Abou-Douh, M. S. Ishak, and Tom J. Mabry. 1994. "Flavonol Glycosides and Phenolics from Withania Somnifera." *Phytochemistry* 37 (4). https://doi.org/10.1016/S0031-9422(00)89563-1.

Khameneh, Bahman, Roudayna Diab, Kiarash Ghazvini, and Bibi Sedigheh Fazly Bazzaz. 2016. "Breakthroughs in Bacterial Resistance Mechanisms and the Potential Ways to Combat Them." *Microbial Pathogenesis*. https://doi.org/10.1016/j.micpath.2016.02.009.

Ko, Wen Chien, David L. Paterson, Anthanasia J. Sagnimeni, Dennis S. Hansen, Anne von Gottberg, Sunita Mohapatra, Jose Maria Casellas, et al., 2002. "Community-Acquired Klebsiella Pneumoniae Bacteremia: Global Differences in Clinical Patterns." *Emerging Infectious Diseases* 8 (2). https://doi.org/10.3201/eid0802.010025.

Kordali, Saban, Recep Kotan, Ahmet Mavi, Ahmet Cakir, Arzu Ala, and Ali Yildirim. 2005. "Determination of the Chemical Composition and Antioxidant Activity of the Essential Oil of Artemisia Dracunculus and of the Antifungal and Antibacterial Activities of Turkish Artemisia Absinthium, A. Dracunculus, Artemisia Santonicum, and Artemisia Spicigera Essential Oils." *Journal of Agricultural and Food Chemistry* 53 (24). https://doi.org/10.1021/jf0516538.

Kulkarni, Shrinivas K., and Ipe Ninan. 1997. "Inhibition of Morphine Tolerance and Dependence by Withania Somnifera in Mice." *Journal of Ethnopharmacology* 57 (3). https://doi.org/10.1016/S0378-8741(97)00064-0.

Kurdekar, Ranjita R., Ganesh R. Hegde, and Satyanarayana S. Hebbar. 2012. "Antimicrobial Efficacy of Bridelia Retusa (Linn.) Spreng. and Asclepias Curassavica Linn." *Indian Journal of Natural Products and Resources* 3 (4).

Lin, Rong Dih, Yi Ping Chin, and Mei Hsien Lee. 2005. "Antimicrobial Activity of Antibiotics in Combination with Natural Flavonoids against Clinical Extended-Spectrum β-Lactamase (ESBL)-Producing Klebsiella Pneumoniae." *Phytotherapy Research* 19 (7). https://doi.org/10.1002/ptr.1695.

Lis-Balchin, M., and S. G. Deans. 1997. "Bioactivity of Selected Plant Essential Oils against Listeria Monocytogenes." *Journal of Applied Microbiology* 82 (6). https://doi.org/10.1046/j.1365-2672.1997.00153.x.

Liu, Chengcheng, Yingli Zhou, Li Wang, Lei Han, Jin'e Lei, Hafiz Muhammad Ishaq, Sean P. Nair, and Jiru Xu. 2016. "Photodynamic Inactivation of Klebsiella Pneumoniae Biofilms and Planktonic Cells by 5-Aminolevulinic Acid and 5-Aminolevulinic Acid Methyl Ester." *Lasers in Medical Science* 31 (3). https://doi.org/10.1007/s10103-016-1891-1.

Lu, Mengjiao, Cuixiang Yu, Xueyan Cui, Jinyi Shi, Lei Yuan, and Shujuan Sun. 2018. "Gentamicin Synergises with Azoles against Drug-Resistant Candida Albicans." *International Journal of Antimicrobial Agents* 51 (1). https://doi.org/10.1016/j.ijantimicag.2017.09.012.

McCabe, Robert, Larry Lambert, and Brad Frazee. 2010. "Invasive Klebsiella Pneumoniae Infections, California, USA." *Emerging Infectious Diseases*. https://doi.org/10.3201/eid1609.100386.

McFadden, David W., Dale R. Riggs, Barbara J. Jackson, and Linda Vona-Davis. 2003. "Keyhole Limpet Hemocyanin, a Novel Immune Stimulant with Promising Anticancer Activity in Barrett's Esophageal Adenocarcinoma." *American Journal of Surgery* 186 (5). https://doi.org/10.1016/j.amjsurg.2003.08.002.

Meatherall, Bonnie L., Daniel Gregson, Terry Ross, Johann D. D. Pitout, and Kevin B. Laupland. 2009. "Incidence, Risk Factors, and Outcomes of Klebsiella Pneumoniae Bacteremia." *American Journal of Medicine* 122 (9). https://doi.org/10.1016/j.amjmed.2009.03.034.

Meletis, Georgios, Egki Tzampaz, Effrosyni Sianou, Ioannis Tzavaras, and Danai Sofianou. 2011. "Colistin Heteroresistance in Carbapenemase-Producing Klebsiella Pneumoniae." *Journal of Antimicrobial Chemotherapy.* https://doi.org/10.1093/jac/dkr007.

Molan, P. C., and J. A. Betts. 2004. "Clinical Usage of Honey as a Wound Dressing: An Update." *Journal of Wound Care.* https://doi.org/10.12968/jowc.2004.13.9.26708.

Molloy, Michael J., Nicolas Bouladoux, and Yasmine Belkaid. 2012. "Intestinal Microbiota: Shaping Local and Systemic Immune Responses." *Seminars in Immunology.* https://doi.org/10.1016/j.smim.2011.11.008.

Nascimento, Gislene G. F., Juliana Locatelli, Paulo C. Freitas, and Giuliana L. Silva. 2000. "Antibacterial Activity of Plant Extracts and Phytochemicals on Antibiotic-Resistant Bacteria." *Brazilian Journal of Microbiology* 31 (4). https://doi.org/10.1590/S1517-83822000000400003.

Noronha C, and Almeida A. 2000. "Local Burn Treatment -Topical Antimicrobial Agents." *Annals of Burns and Fire Disasters.*

Owolabi, Omonkhelin J., Eric K. I. Omogbai, and Osahon Obasuyi. 2007. "Antifungal and Antibacterial Activities of the Ethanolic and Aqueous Extract of Kigelia Africana (Bignoniaceae) Stem Bark." *African Journal of Biotechnology* 6 (14). https://doi.org/10.5897/AJB2007. 000-2244.

Papp-Wallace, Krisztina M., and Robert A. Bonomo. 2016. "New β-Lactamase Inhibitors in the Clinic." *Infectious Disease Clinics of North America.* https://doi.org/10.1016/j.idc.2016.02.007.

Paterson, David L, and Robert A Bonomo. 2005. "Clinical Update Extended-Spectrum Beta-Lactamases : A Clinical Update." *Clinical Microbiology Reviews* 18 (4). https://doi.org/10.1128/CMR.18.4.657.

Paukner, Susanne, Lars Hesse, Andrej Prežeľj, Tomaž Šolmajer, and Uroš Urleb. 2009. "In Vitro Activity of LK-157, a Novel Tricyclic Carbapenem as Broad-Spectrum β-Lactamase Inhibitor." *Antimicrobial*

Agents and Chemotherapy 53 (2). https://doi.org/10.1128/AAC.00085-08.

Podschun, R, and U Ullmann. 1998. "*Klebsiella Spp. as Nosocomial Pathogens: Epidemiology, Taxonomy, Typing Methods, and Pathogenicity Factors.*" Vol. 11. http://cmr.asm.org/.

Poudyal, Anima, Benjamin P. Howden, Jan M. Bell, Wei Gao, Roxanne J. Owen, John D. Turnidge, Roger L. Nation, and Jian Li. 2008. "In Vitro Pharmacodynamics of Colistin against Multidrug-Resistant Klebsiella Pneumoniae." *Journal of Antimicrobial Chemotherapy* 62 (6). https://doi.org/10.1093/jac/dkn425.

Pournaras, Spyros, Georgia Vrioni, Evangelia Neou, John Dendrinos, Evangelia Dimitroulia, Aggeliki Poulou, and Athanassios Tsakris. 2011. "Activity of Tigecycline Alone and in Combination with Colistin and Meropenem against Klebsiella Pneumoniae Carbapenemase (KPC)-Producing Enterobacteriaceae Strains by Time-Kill Assay." *International Journal of Antimicrobial Agents* 37 (3). https://doi.org/10.1016/j.ijantimicag.2010.10.031.

Pradhan, D., P. K. Panda, and G. Tripathy. 2009. "Evaluation of the Immunomodulatory Activity of the Methanolic Extract of Couroupita Guianensis Aubl. Flowers in Rats." *Natural Product Radiance* 8 (1).

Pugh, Nirmal, Samir A. Ross, Mahmoud A. ElSohly, and David S. Pasco. 2001. "Characterization of Aloeride, a New High-Molecular-Weight Polysaccharide from Aloe Vera with Potent Immunostimulatory Activity." *Journal of Agricultural and Food Chemistry* 49 (2). https://doi.org/10.1021/jf001036d.

Qureshi, Zubair A., David L. Paterson, Brian A. Potoski, Mary C. Kilayko, Gabriel Sandovsky, Emilia Sordillo, Bruce Polsky, Jennifer M. Adams-Haduch, and Yohei Doi. 2012. "Treatment Outcome of Bacteremia Due to KPC-Producing Klebsiella Pneumoniae: Superiority of Combination Antimicrobial Regimens." *Antimicrobial Agents and Chemotherapy* 56 (4). https://doi.org/10.1128/AAC.06268-11.

Sader, Helio S., Mariana Castanheira, Robert K. Flamm, Rodrigo E. Mendes, David J. Farrell, and Ronald N. Jones. 2015. "Ceftazidime/Avibactam Tested against Gram-Negative Bacteria from

Intensive Care Unit (ICU) and Non-ICU Patients, Including Those with Ventilator-Associated Pneumonia." *International Journal of Antimicrobial Agents* 46 (1). https://doi.org/10.1016/j.ijantimicag.2015. 02.022.

Sahin, F., M. Güllüce, D. Daferera, A. Sökmen, M. Sökmen, M. Polissiou, G. Agar, and H. Özer. 2004. "Biological Activities of the Essential Oils and Methanol Extract of Origanum Vulgare Ssp. Vulgare in the Eastern Anatolia Region of Turkey." *Food Control* 15 (7): 549–57. https://doi.org/10.1016/j.foodcont.2003.08.009.

Satlin, Michael J., Liang Chen, Gopi Patel, Angela Gomez-Simmonds, Gregory Weston, Angela C. Kim, Susan K. Seo, et al., 2017. "Multicenter Clinical and Molecular Epidemiological Analysis of Bacteremia Due to Carbapenem-Resistant Enterobacteriaceae (CRE) in the CRE Epicenter of the United States." *Antimicrobial Agents and Chemotherapy* 61 (4). https://doi.org/10.1128/AAC.02349-16.

Schentag, Jerome J., and William J. Jusko. 1977. "Renal Clearance and Tissue Accumulation of Gentamicin." *Clinical Pharmacology and Therapeutics* 22 (3). https://doi.org/10.1002/cpt1977223364.

Schliebs, Reinhard, André Liebmann, Salil K. Bhattacharya, Ashok Kumar, Shibnath Ghosal, and Volker Bigl. 1997. "Systemic Administration of Defined Extracts from Withania Somnifera (Indian Ginseng) and Shilajit Differentially Affects Cholinergic but Not Glutamatergic and Gabaergic Markers in Rat Brain." *Neurochemistry International* 30 (2). https://doi.org/10.1016/S0197-0186(96)00025-3.

Shah Pratibha, J., and T. Williamson Manita. 2015. "Antibacterial Activity of Honey against ESBL Producing Klebsiella Pneumoniae from Burn Wound Infections." *International Journal of Current Pharmaceutical Research* 7 (2).

Shields, Ryan K., Brian A. Potoski, Ghady Haidar, Binghua Hao, Yohei Doi, Liang Chen, Ellen G. Press, Barry N. Kreiswirth, Cornelius J. Clancy, and M. Hong Nguyen. 2016. "Clinical Outcomes, Drug Toxicity, and Emergence of Ceftazidime-Avibactam Resistance Among Patients Treated for Carbapenem-Resistant Enterobacteriaceae Infections." *Clinical Infectious Diseases : An Official Publication of the Infectious*

Diseases Society of America 63 (12). https://doi.org/10.1093/cid/ciw636.

Shrivastava, Neeta, Astha Varma, and Harish Padh. 2011. "Andrographolide: A New Plant-Derived Antineoplastic Entity on Horizon." *Evidence-Based Complementary and Alternative Medicine.* https://doi.org/10.1093/ecam/nep135.

Siani, Antonio C., Mônica F. S. Ramos, Anderson C. Guimarães, Glória S. Susunaga, and Maria Das G. B. Zoghbi. 1999. "Volatile Constituents from Oleoresin of Protium Heptaphyllum (Aubl.) March." *Journal of Essential Oil Research* 11 (1). https://doi.org/10.1080/10412905.1999.9701075.

Soares, Adriana C., Danielle G. Souza, Vanessa Pinho, Angélica T. Vieira, Jacques R. Nicoli, Fernando Q. Cunha, Alberto Mantovani, Luiz Fernando L. Reis, Adriana A. M. Dias, and Mauro M. Teixeira. 2006. "Dual Function of the Long Pentraxin PTX3 in Resistance against Pulmonary Infection with Klebsiella Pneumoniae in Transgenic Mice." *Microbes and Infection* 8 (5). https://doi.org/10.1016/j.micinf.2005.12.017.

Souza, T. C., A. M. Silva, J. R. P. Drews, D. A. Gomes, C. G. Vinderola, and J. R. Nicoli. 2013. "In Vitro Evaluation of Bifidobacterium Strains of Human Origin for Potential Use in Probiotic Functional Foods." *Beneficial Microbes* 4 (2). https://doi.org/10.3920/BM2012.0052.

Stachyra, Thérèse, Premavathy Levasseur, Marie Claude Péchereau, Anne Marie Girard, Monique Claudon, Christine Miossec, and Michael T. Black. 2009. "In Vitro Activity of the β-Lactamase Inhibitor NXL104 against KPC-2 Carbapenemase and Enterobacteriaceae Expressing KPC Carbapenemases." *Journal of Antimicrobial Chemotherapy* 64 (2): 326–29. https://doi.org/10.1093/jac/dkp197.

Trabattoni, D., M. Clerici, S. Centanni, M. Mantero, M. Garziano, and F. Blasi. 2017. "Immunomodulatory Effects of Pidotimod in Adults with Community-Acquired Pneumonia Undergoing Standard Antibiotic Therapy." *Pulmonary Pharmacology and Therapeutics* 44. https://doi.org/10.1016/j.pupt.2017.03.005.

Tumbarello, Mario, Enrico Maria Trecarichi, Francesco Giuseppe de Rosa, Maddalena Giannella, Daniele Roberto Giacobbe, Matteo Bassetti, Angela Raffaella Losito, et al., 2015. "Infections Caused by KPC-Producing Klebsiella Pneumoniae: Differences in Therapy and Mortality in a Multicentre Study." *Journal of Antimicrobial Chemotherapy* 70 (7). https://doi.org/10.1093/jac/dkv086.

Tzouvelekis, L. S., A. Markogiannakis, E. Piperaki, M. Souli, and G. L. Daikos. 2014. "Treating Infections Caused by Carbapenemase-Producing Enterobacteriaceae." *Clinical Microbiology and Infection.* https://doi.org/10.1111/1469-0691.12697.

Vatanavicharn, Tipachai, Premruethai Supungul, Narongsak Puanglarp, Wanchart Yingvilasprasert, and Anchalee Tassanakajon. 2009. "Genomic Structure, Expression Pattern and Functional Characterization of CrustinPm5, a Unique Isoform of Crustin from Penaeus Monodon." *Comparative Biochemistry and Physiology - B Biochemistry and Molecular Biology* 153 (3). https://doi.org/10.1016/j.cbpb.2009.03.004.

Verma, Vivek, Kusum Harjai, and Sanjay Chhibber. 2009. "Restricting Ciprofloxacin-Induced Resistant Variant Formation in Biofilm of Klebsiella Pneumoniae B5055 by Complementary Bacteriophage Treatment." *Journal of Antimicrobial Chemotherapy* 64 (6). https://doi.org/10.1093/jac/dkp360.

Wood, Simon, Daniel Metcalf, Deirdre Devine, and Colin Robinson. 2006. "Erythrosine Is a Potential Photosensitizer for the Photodynamic Therapy of Oral Plaque Biofilms." *Journal of Antimicrobial Chemotherapy* 57 (4): 680–84. https://doi.org/10.1093/jac/dkl021.

Xu, Fei, Ying Li, Justen Ahmad, Yonggang Wang, Dorothy Scott, and Jaroslav G Vostal. 2018. "Vitamin K5 Is an Efficient Photosensitizer for Ultraviolet A Light Inactivation of Bacteria." *FEMS Microbiology Letters.* https://doi.org/10.1093/femsle/fny005/4810545.

Xu, J., F. Zhou, B. P. Ji, R. S. Pei, and N. Xu. 2008. "The Antibacterial Mechanism of Carvacrol and Thymol against Escherichia Coli." *Letters in Applied Microbiology* 47 (3): 174–79. https://doi.org/10.1111/j.1472-765X.2008.02407.x.

Zazo, Hinojal, Clara I. Colino, and José M. Lanao. 2016. "Current Applications of Nanoparticles in Infectious Diseases." *Journal of Controlled Release*. Elsevier B. V. https://doi.org/10.1016/j.jconrel.2016.01.008.

Zusman, Oren, Sergey Altunin, Fidi Koppel, Yael Dishon Benattar, Habip Gedik, and Mical Paul. 2017. "Polymyxin Monotherapy or in Combination against Carbapenem-Resistant Bacteria: Systematic Review and Meta-Analysis." *Journal of Antimicrobial Chemotherapy* 72 (1): 29–39. https://doi.org/10.1093/jac/dkw377.

Table 1: Plants having antimicrobial activity against *K. pnuemoniae* (adapted from Shaik,et al., 2014; Sharma et al., 2010; Nimri et al., 1999; Bonjar 2004).

S/No	Family	Plant name	Plant part	Solvent	Strain	Antibiotic Used	Reference
1	Euphobiaceae	Acalypa indica	Leaves	Methanol	MTCC 3384	Chloramphenicol	Arutselvi et al., 2012
2	Asclepiadaceae	Asclepias curassavica Linn.	roots	Chloroform, Ethanol, Acetone and Water	MTCC 109	Streptomycin	Kurdekar et al., 2012
3	Salvadoraceae	Salvadora oleoides Decne	stem	Benzene	MTCC 3384	Streptomycin	Kumar et al., 2012
4	Euphorbiaceae	Phylanthus niruri	Leaves	Methanol	MTCC 3384	Chloramphenicol	Arutselvi et al., 2012
5	Lecythidaceae	Couroupita guianensisAubl	Fruits	Chloroform	MTCC 109; ESBL 75799; ESBL 3971; ESBL 3967; ESBL 3894	Streptomycin	Rauf et al., 2012
6	Solanacaea	Datura stramonium	stem-bark	Ethanolic	KP72011FMC	Gentamycin	Shagal et al., 2012
7	Asclepiadaceae	Gymnema sylvestre (Retz) R. Br ex. Schultes	Leaf(L) and stem(S)	Chloroform, Acetone, Petroleum Ether, Water, Methanol,	MTCC 109	chloramphenicol and tetracycline	Murugan et al., 2012

Table 1. (Continued)

S/No	Family	Plant name	Solvent	Plant part	Strain	Antibiotic Used	Reference
8	Polygonaceae	Persicaria piripu (DC.) M.R.Almeida	Ethanol and Water, Chloroform, Methanolic, Acetone,	Leaves	MTCC 109	Streptomycin	Kurdekar et al., 2012
9	Cupressaceae	Cupressus sempervirens L	Ethyl Acetate and Ethanolic	aerial parts	MTCC 618	Pencillin	Chaudhary et al., 2012
10	Mimosaceae	Acacia mearnsii De Wild.	Methanol	bark	ATCC 10031	Tetracycline	Olajuyigbe et al., 2012
11	Lobeliaceae	Lobelia nicotianaefolia Roth ex R. & S	Acetone, Chloroform, Water and Ethanol	Leaves and roots	MTCC 109	Streptomycin	Kurdekar et al., 2012
12	Moraceae	Ficus benghalensis Linn.	Chloroform, Methanol	leaves	MTCC B2405	Ciproflaxacin	Koon and Rao, 2012

Table. 2: Essential oil producing plants showing inhibitory action against *K. pnuemoniae* (adapted from Shaik et al., 2014; Xhaxhiu et al., 2013, Fournomiti et al., 2015).

S/No	Plant Part	Strain	Plant Name	Reference
1	Leaves	MTCC 4030	Eucalyptus globulus	Bharti et al., 2012
2	Leaves	PTCC-1053	Cinnamomum zeylanicum	Hadi et al., 2012
3	Flower	PTCC 1290	Allium rotundum	Dehpour et al., 2012
4	Flowering tips and Leaves	MTCC 4030	Thymus vulgaris	Bharti et al., 2012
5	Fruits	MTCC 4030	Cinnamomum cecidodaphne	Bharti et al., 2012
6	Aerial parts	CIP 106818	Matricaria pubescens	Makhloufi et al., 2012
7	Leaves	MTCC 4030	Zanthoxylum rhetsa	Bharti et al., 2012
8	Aerial parts	WHO24	Rosmarinus officinalis	Chobba et al., 2012
9	Seeds	MTCC 4030	Coriandrum sativum	Bharti et al., 2012
10	Seeds	----	Sphallerocarpus gracilis	Bernardes et al., 2010
11	Seedlings	----	Origanum vulgare	Ozkalp et al., 2010

Table 3: Management of *K. pnumoniae* infections via phages (adapted from Cooper et al., 2018; Morozova et al., 2018;

S No.	Combination	Bacterial Target	Results	References
1	Phage B5055/CIP	*K. pnuemoniae*	Resistant variants frequency was reduced within combined testing as compared with individual components. When phage was added so it caused almost 5log10 reduction in biofilm of bacteria in 180 min afterward the addition of phage..	Verma et al., 2009
2	Phage KPO1K2 or NDP/CoSO4 or FeCl3	*K. pneumoniae*	$\geq$5log10 reduction within KPO1K2/10 µM FeCl3 + 500 µM CoSO4 combination in 3 day old biofilms versus untreated control. Reduction of $\leq$1log10 in NDP/10 µM FeCl3 combination in biofilms up to seven days versus untreated control. Reduction of 1–2log10 in KPO1K2/10 µM FeCl3 combination in biofilms up to seven days versus untreated control.	Chhibber et al., 2013
3	Phage Kpn5/HPMC hydrogel	*K pneumoniae*	When phage was applied so it increased the survival ($\geq$60%) of mice which was burnt over five days	Kumari et al., 2011
4	K5-2 and K5-4	*Klebsiella*	K5-2 produces spots on 7 capsular kinds of Klebsiella. When phage K5-4 was used so it enhanced the mice survival which was treated with strain of K. pneumoniae K5. Each type of bacteriophage encodes for 2 diverse capsule depolymerases which enables them to replicate on certain strains of Klebsiella	Hsieh et al., 2017

5	φK64-1	*Klebsiella*	Eight putative depolymerases were encoded by the phage. Phage mutants production which did not encode putative depolymerase's eradicated the lytic action.	Pan et al., 2017
S No.	Combination	Bacterial Target	Results	References
6	KP32	*Klebsiella*	TTPA (i-e Tail tubular protein A), which is the structural protein found in tail region of KP32, shows lytic action against EPS. TTPA is a macromolecule which has dual functions like it possess both enzymatic and structural activities	Pyra et al., 2017

Table 4: Antimicrobial agents towards KPC-KP (adapted from Bassetti et al., 2018; Morrill et al., 2015)

S No.	Loading dose	Drug	Comments	Daily dose for normal renal function
1	1-2 g	Meropenem	Meropenem must be applied along with another active substance; the likelihood of response is greater when MIC of meropenem is about 8 mg/L. Salvage remedy in association with two types of carbapenems, for example ertapenem in combination with either doripenem or meropenem can be considered when further treatment options are not available or suitable.	2 g every 8 hours IV infused over 3-6 hours
2	100-200 mg	Tigecycline	For pneumonia or BSIs, tigecycline MIC >0.5 mg/L, high dosages are suggested (loading dosage, 200 mg is followed via 100 mg each twelve hours), along with other agent. It must not be applied within UTI (urinary tract infections).	50-100 mg every 12 hours IV

Table 4. (Continued)

S No.	Loading dose	Drug	Comments	Daily dose for normal renal function
3	9 million IU	Polymyxins Colistin A	For infections produced via the microbes having MIC >0.5 mg/L, it is recommended to apply colistin as a combination remedy part. For adjustment of dose within patients having renal failure (Nation et al.,2014)	4.5 million IU IV every 12 hours Intrathecal/intraventricular: 125 000 -50 000 IU Inhaled: 1 to 3 million IU every 8 hours
4	Not required	Fosfomycin	Fosfomycin can be applied within combination therapy for infections of KPC-KP and it is given to patients as 6-8 gram each 8 hours. During treatment there are chances of resistance and it must be monitored while the treatment process.	18 to 24 g IV in 3 to 4 doses
5	Not required	Polymyxin B	No adjustment of dosage for kidney failure.	7500-12 500 IU/kg every 12 hours every 12 hours Intrathecal/intraventricular: 50 000 IU every 24 hours
6	Not required	Ceftazidime/avibactam	It is suggested for Ventilator and Hospital acquired pneumonia, severe type of urinary tract and intra-abdominal infections and for treating the infections which are caused by aerobic gram-negative microbes within adult individuals when further treatments options are not effective; invitro it is active towards KPC, Enterobacteriaceae-producing ESBLs, OXA-48	2.5 g every 8 hours IV infused over 2 hours

		and AmpC. Ceftazidime/ avibactam along with another agents is used for the treatment of infections caused by KPC-KP, whether it must be applied in combination or separately remain unobvious, and needs further study.
7	Aminoglycosides	Aminoglycosides could be used as combination therapy part in order to treat infections of KPC-KP, particularly if there is resistance of colistin.

KPC-KP, *Klebsiella pneumoniae* carbapenemase-producing *K. pneumoniae*; BSI, bloodstream infection; MIC, minimum inhibitory concentration; ESBL, extended-spectrum b-lactamase; 1 mg of colistin A base activity is contained within 2.4 mg colistimethate, that is equal to 30 000 IU, while 1 mg polymyxin B is equal to 10 000 IU.

Chapter 3

ANTIBIOTIC RESISTANCE IN *KLEBSIELLA PNEUMONIAE*

Suganthi Rasangam[1], Aris Chandran Abdullah[2] and V. Gopalakrishnan[2,]*

[1]UniKL Royal College of Medicine Perak,
3 Jalan Greentown, Ipoh, Malaysia
[2]Associate Professor,
UniKL Royal College of Medicine Perak,
3 Jalan Greentown, Ipoh, Malaysia

ABSTRACT

Klebsiella pneumoniae is implicated in numerous healthcare associated infections. It commonly causes pneumonia and bacteremia among patients warded in intensive care unit (ICU), newborn unit and immunocompromised individuals. *Klebsiella* species survive in moist environmental sites in hospitals and colonize human bowel, upper respiratory tract, bladder, and skin. Antibiotic resistance among *Klebsiella* species is not only a problem for the patient infected but the whole

---

[*] Corresponding Author's Email: gopalakrishnan@unikl.edu.my.

institution, because of its impact on increased unresponsiveness to treatment. Patients infected with *K. pneumoniae* harboring resistant genes are more difficult to be treated as they are resistant to almost all the antibiotic groups including last resort drug, carbapenems.

These pathogens possess several mechanisms of resistance. The most important resistant genes associated with *K. pneumoniae* are NDM, KPC, VIM, IMP and OXA-48, and each of these belonging to a different class. Ambler class B, Verona imipenemase (VIM), imipenemase (IMP), New Delhi metallo-β-lactamase (NDM) and class D OXA-48 β lactamase are metallo-β-lactamases. Ambler class A enzymes are clavulanic-acid-inhibited are chromosome-encoded enzymes and plasmid-mediated enzymes such as *K. pneumoniae* carbapenemase (KPC) enzymes. Each of these genes express different type of resistance towards antibiotics.

K. pneumoniae which harbor antibiotic resistant genes have spread worldwide, and hence make it a potential threat to currently available dwindling antibiotic-based remedies. Carbapenems such as imipenem, meropenem and ertapenem which are commonly used to treat patients have become ineffective for treating patients. Identification and detection of *K. pneumoniae* which are resistant to carbapenem group of antibiotics maybe difficult based on routine microbiological procedures and may limit the efficacy of infection control measures implemented in healthcare setups. Effective and rapid techniques through a multipronged approach to detect these resistant genes by molecular methods (PCR, multiplex PCR, etc.) are the need of the hour to overcome the outcome of antibiotic resistance.

Keywords: Antibiotic Resistance, Klebsiella, Hydrolyzing enzymes, Efflux pump

1. INTRODUCTION

Antibiotics are compounds that have been discovered, commercialized, and administered routinely to treat infections. They are an important medical intervention needed for treatment against bacterial infections, complex medical approaches, complicated surgical procedure, organ transplantation and cancer management. Lately, unfortunately, there is marked increase in antibiotic resistance among bacterial pathogens, particularly isolates of *K. pneumoniae*. Compared to susceptible strains, infections caused by multidrug-resistance (MDR) organisms have increased the morbidity and mortality (Cosgrove, S. E. 2006; Diaz et al. 2005; Thomas et al. 2005).

To understand the problem of antibiotic resistance, it is essential to discuss some relevant concepts. Firstly, most of the antibiotics are naturally produced moieties and thus, co-resident bacteria have developed strategies to overcome their action to survive. They become 'intrinsically' resistant to one or more antibiotics. But bacteria having intrinsic factors of resistance are not the point of concern. We should be aware or be concerned about 'acquired resistance' in bacterial population that was previously vulnerable to these antibiotic compounds. Acquired resistance develop due to mutations in chromosomal genes or acquiring of external genetic determinants of resistance which will be discussed later in the chapter. Secondly, breakpoints for antibiotic compounds are established on in-vitro studies of antibiotics against substantial number of bacterial isolates. The elucidation of these susceptibility patterns may vary according to clinical scenarios and place of occurrence. In contrast, in vivo susceptibility of an organism to an antibiotic may vary as well. This chapter will briefly outline and focus on molecular, biochemical and other mechanisms of bacterial resistance subsequently.

2. Resistance Mechanisms

Resistance against a wide range of antibiotics including carbapenems is common among *K. pneumoniae* strains of recent. *K pneumoniae* produce extended spectrum-β-lactamases (ESBL), metallo-β-lactamases (MBLs), carbapenemases, *Klebsiella pneumoniae* carbapenemase (KPC), New Delhi metallo-β-lactamases (NDM), imipenem-hydrolyzing enzyme (IMP), Verona integron-encoded metallo β-lactamase (VIM) and OXA-48(11). *K pneumoniae* are intrinsically resistant to ampicillin and lately have acquired resistant genes mainly through plasmids and chromosomes. There are many ways that *K. pneumoniae* acquire these resistant genes viz., transformation, conjugation, and transduction in which susceptible strains may acquire resistant genes from other bacteria and sources. At present, there are a few mechanisms that *K pneumoniae* acquire antibiotic resistant genes. They are target alteration, changes in permeability of the membrane, efflux pumps and enzyme degradation of antibiotics (Bratu et al. 2005).

Mutational resistance is a group of cells derived from a vulnerable group that develop mutation in genes that will alter the action of the drug on it. Once a resistant mutant emerges, the antibiotic will eliminate the susceptible portion and resistant bacteria will predominate. Mutations resulting in antimicrobial resistance will alter the antibiotic action through one of the following mechanisms. They are modification of the antibiotic target, decrease in the drug intake, and activation of efflux mechanism to expel the harmful molecule in important metabolic pathways by modulating the regulatory mechanisms (Winn et al. 2006).

Horizontal gene transfer is by acquisition from foreign DNA that is a frequent cause for the occurrence of antibiotic resistance. External genetic material are acquired by the bacteria by transformation, conjugation and transduction. Simplest type of horizontal gene transfer is transformation, which is observed only in a few bacterial species that are capable of 'naturally' incorporating naked DNA to develop resistance. Conjugation is the major mode acquiring resistance in hospital environment and is a very efficient method of gene transfer, which involves cell-to-cell contact. It occurs rapidly in the gastrointestinal tract of humans who take antibiotics. Conjugation also uses mobile genetic elements (MGE) (Nordmann et al. 2011). Plasmids and transposons are most important MGE. Finally, integrons are one of the most efficient mechanisms for acquiring antibiotic resistance (Wilson et al. 2014).

Target alterations or modifications of the antibiotic molecule occur by enzymes produced by the organism that inactivates the drug by the addition of specific chemicals or by destroying the molecule itself and making the antibiotic incapable to interact with its target (Winn et al. 2006). Another method involved in this mechanism is by destructing the antibiotic molecule which is the chief mechanism of β-lactam resistance that relies on the destruction of the compound by β-lactamases. The enzyme, β-lactamase was first described in early 1940s (Abraham and Chain 2010; D'Costa et al. 2011). A new plasmid-encoded β-lactamase that had the capability to hydrolyze ampicillin was identified among gram-negative organisms (TEM-1) during the 1960s (Sydnor and Perl 2011). Genes that encode for β-lactamases are termed as *bla,* followed by the name of specific enzyme (e.g., *bla* KPC) and have been identified in the chromosome or localized in the mobile genetic elements.

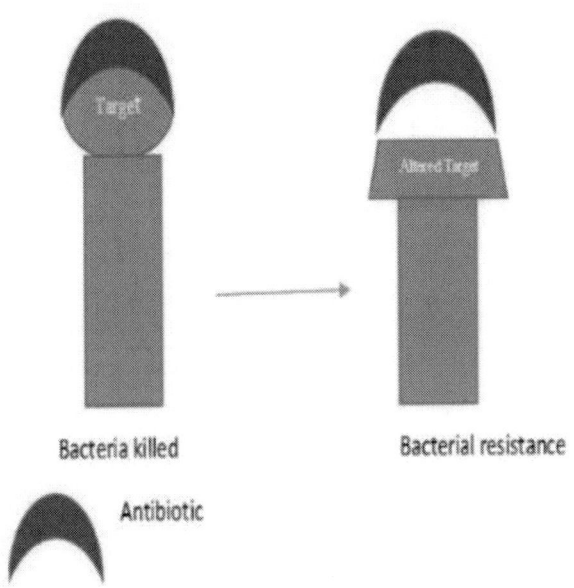

Figure 1. Target alterations or modifications of the antibiotic molecule occur by enzymes produced be the organisms that inactivates the drug by the addition if specific chemicals or by destroying the molecule itself making the antibiotic incapable to interact with its target (Winn et al. 2006). Another method involved in this mechanism is by destructing the antibiotic molecule which is the chief mechanism of β-lactam resistance that relies on the destruction of the compound by β-lactamases. The enzyme, β-lactamase was first described in early 1940s (Abraham and Chain 2010, D'Costa et al. 2011). A new plasmid-encoded β-lactamase that had the capability to hydrolyze ampicillin was identified among gram-negative organisms (TEM-1) during the 1960s (Sydnor and Perl 2011). Genes that encode for β-lactamases are termed as *bla*, followed by the name of specific enzyme (e.g., *bla* KPC) and have been identified in the chromosome or localized in the mobile genetic elements. There are two main classification schemes proposed to group this large number of enzymes. The most common grouping used globally is Ambler classification that separates β-lactamases into 4 groups (A, B, C and D) (Ambler 1980).

There are two main classification schemes proposed to group this large number of enzymes. The most common grouping used globally is Ambler classification that separates β-lactamases into 4 groups (A, B, C and D) (Ambler 1980).

3. BACTERIAL ENZYMES IN RESISTANCE

Many types or classes are involved in initiating antimicrobial resistance as outlined in table 1. An insight to the enzymes is enlisted.

Most of the class A enzymes are rendered inactive by clavulanic acid but not cephamycin (cefoxitin and cefotetan).

Class A includes extended spectrum β-lactamases (ESBL) and *K. pneumoniae* carbapenemase (KPC) that are currently prevalent in *K. pneumoniae* is high (Piddock 2006).

Class B enzymes are recognized as metallo- β-lactamases due to their ability to utilize metal ion (usually Zinc) as a cofactor to act upon the β-lactam ring. They act on a wide range of β-lactams, including carbapenems. To date, there are 10 different metallo-carbapenemases. Most of the clinically important class B enzymes belong to IMP, VIM and NDM families.

The first case reported for IMP-type enzymes was in Japan in the 1990s in *Serratia marcescens,* and from thereon more than 20 different subtypes have been reported to occur throughout the world. The first instance of VIM-type enzymes was described in later part of 1990 in Verona, Italy which has spread worldwide since then. A different carbapenemase was identified in *K. pneumoniae* isolate in 2008 from a Swedish patient who had been previously hospitalized in New Delhi, India. In reference to its origin of the enzyme in New Delhi it was named NDM-1. Emergence of NDM-1 was so alarming and concerning because the *bla* NDM gene was found to be transmissible among different types of gram-negative organisms and spreading globally in a very short span of time, and thus became the most dreaded resistance determinant throughout the world (Cornaglia et al. 2011).

Table 1. Ambler class classification for carbapenemases

Carbapenemase	Ambler Class	Comments	Susceptibility Pattern
KPC *Klebsiella pneumoniae* carbapenemase	Class A	KPC enzymes, plasmid transmission commonest CRE in the U.S.A.	R: AMX, TZP, CTX, CAZ ATM I/R: ETP S/I/R: IMP MER
IMP-carbapenemases (metallo-β-lactamase)	Class B	Identified in Japan in 1990s in *Enterobacteriaceae,* Acinetobacter spp *Pseudomonas*	R: AMX, AMC, CTX, I/R: TZP, CAZ, ETP S/I/R: IMP, MER S: ATM
VIM (Verona integron-encoded metallo-β-lactamase)	Class B	Occur evenly in North and South-America, Europe and Far East. Predominant in *P. aeruginosa* Rare in *Enterobacteriaceae*	R: AMX, AMC, CTX, I/R: TZP, CAZ, ETP S/I/R: IMP, MER S: ATM
NDM-1 (New Delhi metallo-β-lactamase)	Class B	Identified in New Delhi 2009	R: AMX, AMC, CTX, I/R: TZP, CAZ, ETP S/I/R: IMP, MER S: ATM

Table 1. (Continued)

Carbapenemase	Ambler Class	Comments	Susceptibility Pattern
OXA (oxacillinase)	Class D	Occur mainly in *Acinetobacter* spp.	R: AMX, AMC S/I: CTX, IMP, ETP, MER S: CAZ, ATM

R: Resistant; S: Susceptible; I: Intermediate; AMX: amoxicillin; AMC: amoxicillin–clavulanic acid; TZP: piperacillin–tazobactam; CTX: cefotaxime; CAZ: ceftazidime; IMP: imipenem; ETP: ertapenem; MER: meropenem; ATM: Aztreonam (Nordmann, Gniadkowski, Giske, & Po, 2012).

Class C β-lactamases are resistant to all penicillin and cephalosporin group of drugss that do not hydrolyze aztreonam and are not inhibited by clavulanic acid. Amp C is the most relevant class C enzyme in clinical setup (Jacoby 2009).

Class D β-lactamases enzymes (OXA) belong to the class A penicillinases due to their ability to act on oxacillin (hence acquires that name).

Many types of OXA have been reported, for instance OXA-48 is a widely reported class D carbapenemases was initially reported in Turkey in 2001 (Paterson and Bonomo 2005) from a *K pneumoniae* multidrug resistant isolate.

4. Efflux Pumps

Decreased permeability is a process wherein bacteria develop an ability to prevent the antibiotic in reaching its intracellular or periplasmic site located in the cytoplasmic membrane (inner membrane) by decreasing the uptake of antibiotic molecule. Gram-negative bacteria capable of producing complex bacterial machineries can extrude a toxic compound out of the cell that results in antibiotic resistance, and often referred to as efflux pumps (Figure 2).

In other words, bacteria can pump out the antibiotic substance out of the cytoplasm and this mechanism maybe substrate-specific or have broad substrate specificity, which were commonly found in multi-drug resistant (MDR) bacteria (Poole 2005).

5. COMMON STRATEGIES IN GRAM-NEGATIVE BACTERIA

Gram-negative bacteria also have a common strategy to develop antibiotic resistance, which avoid the activity of the antibiotics by interfering with their target sites. Bacteria employ different strategies to achieve this.

Some Gram-negative bacteria are able to protect the target site located in their chromosome while some are capable to modify the target site, which is one of the most common mechanism of resistance to antibiotics in bacteria. Regardless of the type of change (point mutation, enzymatic alteration, replacement, or avoidance of original target), the final outcome is decreased affinity of the antibiotics to the target site (Manson et al. 2010).

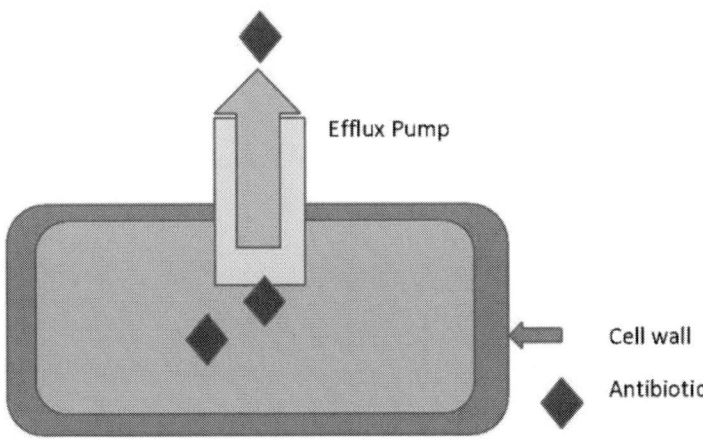

Figure 2. Efflux Pump.

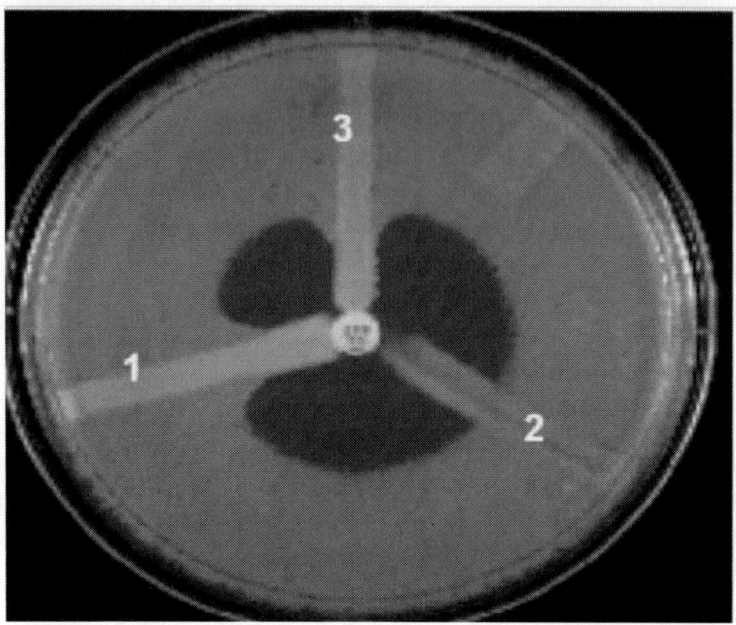

Figure 3. After 16-24 hours of incubation, examine plate for a clover leaf-type indentation at the intersection of the test organism and the *E. coli* 25922, within the zone of inhibition of the Meropenem disk. 1. Positive, 2. Negative, 3. Positive.

6. CONVENTIONAL METHODS IN CARBAPENEMASE DETECTION

Currently, there are a few methods be used in microbiology laboratory to detect carbapenemases among *K. pneumoniae* isolates that show reduced susceptibility (intermediate/resistant) towards carbapenem group of antibiotics and should be tested further. Currently, two methods are being used, Modified Hodge test and modified carbapenem inactivation method.

Culture-based detection is always cost effective but is time consuming, and sensitivity may not be at desirable levels. Moreover, culture-based methods may be unable to identify an isolate as a carbapenemase producing strain.

Polymerase chain reaction (PCR) is relatively simple and is widely used. This requires a shorter duration than conventional techniques to detect antibiotic resistance.

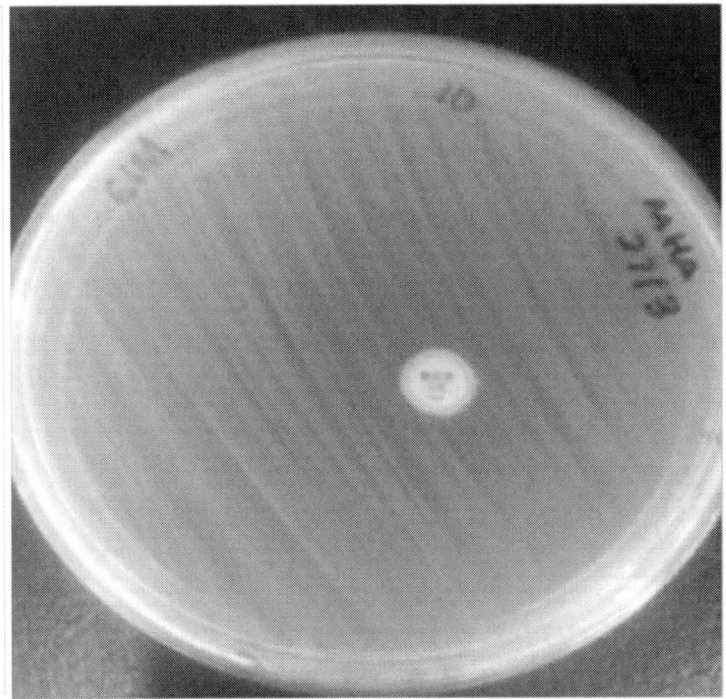

Figure 4. Modified Carbapenem inactivation. Zone diameter of 6-15mm is Carbapenemase positive. Zone diameter of ≥19mm Carbapenemase negative.

It is highly sensitive and requires only a few organisms (alive or dead) for detection and amplification of the specific sequences. The denaturation, annealing and elongation process carried over a series of temperatures is considered as a single amplification cycle. Each step of the cycle should be optimized for the template with specific primer sets. There are forward and reverse primers for PCR reaction each of the sequences to be amplified. Of the different formats of PCR, real time PCR is very useful and is also quantitative. Real-time PCR can be a good choice for rapid method of resistance gene detection and can detect more than one carbapenemase genes from isolated colonies.

Avenues to prevent antibiotic resistance should be addressed by multipronged approach like, improving antibiotics prescribing/stewardship, compliance by the patients, discovery of newer antibiotics, and basic methods of prevention of acquiring organisms by good hygiene, safe food preparation, immunization etc. Antibiotic resistant organisms if detected in hospitals calls for greater infection control approaches and needs an elaborate article by itself.

7. CONCLUSION

Emergence of antibiotic resistance has rapidly advanced in the last few decades. It has now become one of the greatest public health concern globally. This threat has worsened by enormity of resistance against all classes of antibiotics, and on top of it the slower discovery of newer classes and shortage of antibiotics is on the increase. A whole new and comprehensive understanding is crucial to recognize and identify the mechanisms by which these bacteria become resistant to antibiotics.

REFERENCES

Abraham, Edward P. and Chain, Ernst. "An enzyme from bacteria able to destroy penicillin". *Nature,* 146, no. 3713 (1940): 837 - 837.

Ambler, R. P. 1980. "The structure of β-lactamases". *Philos. Trans. R Soc. London B Biol. Sci.*, doi:https://doi.org/10.1098/rstb.1980.0049.

Bratu, S., Tolaney, P., Karumudi, U. 2005. "Carbapenemase-producing *Klebsiella pneumoniae* in Brooklyn, New York: Molecular epidemiology". *J. Antimicrob. Chemother.,* doi: 10.1093/jac/dki175.

Cosgrove, S. E. 2006. "The relationship between antimicrobial resistance and patient outcomes: mortality, length of hospital stay, and health care costs". *Clin. Infect. Dis.,* doi: 10.1086/499406.

Cornaglia, G., Giamarellou, H., Rossolini, G. M. 2011. "Metallo-β-lactamases: a last frontier for β-lactams". *Lancet Infect. Dis.,* doi: 10. 1086/499406.

Diaz Granados, C. A., Zimmer, S. M., Klein, M., Jernigan, J. A. 2005. Comparison of mortality associated with vancomycin-resistant and vancomycin-susceptible enterococcal bloodstream infections: a meta-analysis. *Clin. Infect. Dis.,* doi:https://doi.org/10.1086/430909.

D'Costa, V. M., King, C. E., Kalan, L., Morar, M., Sung, W. W., Schwarz, C., Froese, D., Zazula, G., Calmels, F., Debruyne, R., Golding, G. B., Poinar, H. N., Wright, G. D. 2011. "Antibiotic resistance is ancient". *Nature,* doi:10.1038/nature10388.

Gasink, L. B., Edelstein, P. H., Lautenbach, E. 2009. "Risk factors and clinical impact of *Klebsiella pneumoniae* carbapenemase-producing *Klebsiella pneumoniae*". *Infect. Control Hosp. Epidemiol.,* doi: 10. 1086/648451.

Jacoby, G. A. 2009. "AmpC β-lactamases". *Clin. Microbiol. Rev.,* doi: 10.1128/CMR.00036-08.

Lambert, P. A. 2005. "Bacterial resistance to antibiotics: modified target sites". *Adv. Drug Deliv. Rev.,* doi: 10.1016/j.addr.2005.04.003.

Manson, J. M., Hancock, L. E., Gilmore, M. S. 2010. "Mechanism of chromosomal transfer of *Enterococcus faecalis* pathogenicity island, capsule, antimicrobial resistance, and other traits". *Proc. Natl. Acad. Sci. U S A,* doi: 10.1073/pnas.1000139107.

Nordmann, P., Naas, T., Poirel, L. 2011. "Global spread of carbapenemase-producing Enterobacteriaceae". *Emerg. Infect. Dis.,* doi: 10.3201/ eid1710.110655.

Piddock, L. J. 2006. "Clinically relevant chromosomally encoded multidrug resistance efflux pumps in bacteria". *Clin. Microbiol. Rev.,* doi: 10. 1128/CMR.19.2.382-402.2006.

Poole, K. 2005. "Efflux-mediated antimicrobial resistance". *J. Antimicrob. Chemother.,* doi: 10.1093/jac/dki171.

Paterson, D. L., Bonomo, R. A. 2005. "Extended-spectrum beta-lactamases: a clinical update". *Clin. Microbiol. Rev.,* doi: 10.1128/CMR.18.4.657-686.2005.

Sydnor, E. R., Perl, T. M. 2011. "Hospital epidemiology and infection control in acute-care settings". *Clin. Microbiol. Rev.,* doi: 10.1128/CMR.00027-10.

Thomas, C. M., Nielsen, K. M. 2005. "Mechanism of and barriers to, horizontal gene transfer between bacteria". *Nat. Rev. Microbiol.,* doi: 10.1038/nrmicro1234.

Wilson, D. N. 2014. "Ribosome-targeting antibiotics and mechanisms of bacterial resistance". *Nat. Rev. Microbiol.,* doi: 10.1038/nrmicro3155.

Winn, Allen, Janda, Koneman. 2006. "*Koneman's Color Atlas and Textbook of Diagnostic Microbiology*; 6th Edition".

In: Recent Trends in Understanding …   ISBN: 978-1-53618-503-4
Editors: J. Ramakrishnan et al.   © 2020 Nova Science Publishers, Inc.

Chapter 4

MACROPHAGE IMMUNOMODULATION BY BIOMOLECULES FOR ERADICATING *PSEUDOMONAS AERUGINOSA*

R. Sangeetha[1], D. Parimalanandhini[1], K. Mahalakshmi[1], M. Livya Catherene[1], M. Beulaja[2], R. Thiagarajan[3], S. Janarthanan[1], M. Arumugam[1] and R. Manikandan[1,*]

[1]Department of Zoology, University of Madras,
Guindy Campus, Chennai, India
[2]Department of Biochemistry,
Annai Veilankanni's College for Women, Chennai, India
[3]Department of Advanced Zoology and Biotechnology,
Ramakrishna Mission Vivekananda College, Chennai, India

ABSTRACT

Pseudomonas aeruginosa is a prominent ubiquitous Gram-negative bacteria in infections, predominantly among immune compromised

---

* Corresponding Author's Email: manikandanramar@yahoo.co.in.

individuals and people suffering with ventilator-associated pneumonia. Due to their hyper virulence and recalcitrant multi drug resistance to various antibiotics, phage therapies as well as enzymatic eradication methods, it is chiefly difficult to treat, thereby resulting in increased mortality and morbidity worldwide. Since chemical drugs has very less impact on this bacteria, it is essential to strategize alternative therapies using natural biomolecules.

Biomolecules derived from varied sources greatly involve in modulating the immune function of macrophages as they play a crucial role in host defense via innate immunity through receptor-mediated ingestion and thus modulation of phagocytosis by biomolecular factors provide theoretical support for its therapeutic strategies. In this review, we highlight the phagocytic response of macrophages against the *P. aeruginosa* along with its host evasion mechanism, emphasizing the role of biomolecules as effectual immunomodulators.

Keywords: *Pseudomonas aeruginosa*, biomolecules, macrophages, immunomodulation, phagocytosis

1. Introduction

1.1. *Pseudomonas aeruginosa*

The major task of healthcare professionals nowadays is to treat infections caused by Gram-negative bacteria due to their increasing anti-microbial resistance. *Pseudomonas aeruginosa*, a Gram-negative bacterium has a leading role in causing infections, especially in critically ill and immunocompromised patients (Bassetti et al., 2018). It is a nosocomial contaminant that causes about 10-20% infections in hospitalized patients. It may lead to serious infections such as malignant external otitis, endophthalmitis, endocarditis, meningitis, pneumonia and septicemia (Bodey et al., 1983). It produces a large number of multifactorial extracellular products which may contribute to its virulence. The use of chemical antibiotics like gentamicin, carbenicillin and colistin has less effect in preventing bacteremia. Therefore, immunologic prophylaxis and therapies using biomolecules are needed for its treatment.

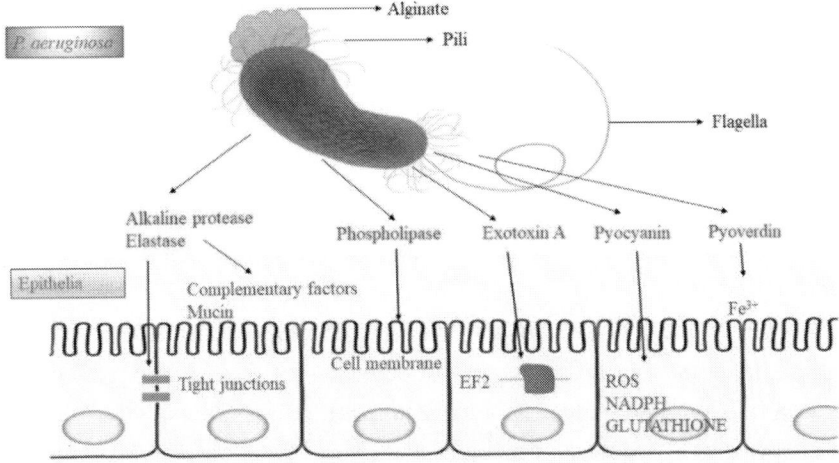

Figure 1. Virulence factors produced by *P. aeruginosa*. The surface appendages, flagella and pili are main adhesins capable of binding to host epithelial gangliosides and they are highly inflammatory. Multiple virulence factors of *P. aeruginosa* have variable effects on the host cell. Proteases such as alkaline protease and elastase have the ability to degrade complement factors, mucins, and disrupt tight junctions between epithelial cells causing the dissemination of bacteria. Lipases and phospholipases target lipids in the surfactant and host cell membranes. Exotoxin A inhibits the elongation factor (EF2) by ADP-ribosylation. Pyocyanin is a blue-green pigments that interferes with electron transport pathways and redox cycling. Pyoverdine arrests Fe^{3+} ions providing a scarcity for free ions in the environment.

1.2. Biomolecules in Inflammation

Biomolecules are biological substances produced by living organisms performing a wide array of functions. The four major biomolecules are carbohydrates, lipids, nucleic acids and proteins which have huge potential in antimicrobial applications (Rogers, 2020). Traditionally, certain biomolecules are being used for the treatment of several pathogenic diseases due to their medicinal properties (Mukopadhyay et al., 2012). Biomolecules such as anti-microbial peptides are being considered as a host defence weapon against microbial pathogens (Rizza et al., 2012). Antibacterial agents selectively destroy bacteria by trespassing their growth and survival using their inherent biochemical as well as biophysical properties. They have an advantage over chemical drugs by having a less cytotoxic effect and possess environmental friendliness (Li et al., 2020). They are involved in modulating the immune system by activating or restoring antimicrobial functions of immune cells like macrophages (Chouhan et al., 2014).

1.3. Macrophage and Its Role in Eradicating This Pathogen

Several evidences support the critical importance of innate immunity, specifically the role of phagocytic cells and their ability to control *P. aeruginosa* infection. Macrophages are professional phagocytic cells that are distributed throughout mammalian organs. They can be activated in response to various types of stimuli produced by infected cells resulting in phagocytosis with hydrolytic enzymes (Elhelu, 1983). In a previous study, nosocomial pneumonia has been treated by surfactant protein D (SP-D) stimulated alveolar macrophages through enhancement of phagocytosis leading to control of pathogenesis (Restrepo et al., 1999). A recent *in vivo* study showed that galactosylceramide activated CD1d-restricted T cells in mice increased the amount of interferon-γ leading to enhanced phagocytosis of *P. aeruginosa* by alveolar macrophages (Nieuwenhuis et al., 2002). Studies have been performed to understand the possible role of macrophages in bactericidal activity against *P. aeruginosa*. In an *in vitro* study, 14C-labeled pseudomonas IgG opsonized bacteria enhanced the uptake of bacteria by *in vitro* cultured alveolar macrophages making it an important respiratory antibody for bacterial infections (Reynolds et al., 1975). This review will spotlight the current knowledge in phagocytic recognition and clearance of bacteria. This will also highlight the context of phagocytic receptors and pathways by immunomodulatory actions of biomolecules. The bacterial strategies for phagocytic avoidance and inflammatory mechanisms that alter the phagocytic clearance of bacteria will be highlighted.

2. BACTERIAL PATHOGENESIS VS PHAGOCYTOSIS

2.1. Bacterial Invasion

P. aeruginosa can invade and replicate inside the host cell within vacuoles or in cytoplasm which is a virulence property of bacteria (Finlay et al., 1988) (Simon et al., 1992). Invasion of the bacteria generally involves the adherence to the host cell surface, active metabolism and evasion of the host defence system leading to changes in the host cell (Fleizszig et al., 1995). Cytotoxic P. aeruginosa causes necrosis of alveolar epithelium which serves as a barrier for the prevention of dissemination of inhaled organisms into circulation from lung, followed by the production of type III secreted toxins leading to acute epithelial injury and bacterial dissemination (Kurahashi et al., 1999). It has increasing antimicrobial resistance which includes multidrug efflux pumps, β-lactamases, down-regulation of outer membrane porins and multidrug resistance (Driscoll et al., 2007). Thus, the effective treatment would be the source control measures using activating our immune system through several anti-pseudomonal biomolecules.

2.2. Host Phagocytosis

P. aeruginosa is found to be effectively opsonized with C3b by the classical complement pathway and to a lesser extent by an alternate pathway (Meshulam et al., 1982). Role of LPS in opsonisation and phagocytosis has been studied in which the structure of O-antigen polysaccharide chain of LPS is found to play a crucial role (Engels et al., 1985). Unopsonized *P. aeruginosa* is phagocytized by human monocyte-derived macrophages which are facilitated by mannose receptors (Speert et al., 1988). Phagocytosis of this bacteria is facilitated by the presence of many pathogen-associated molecular patterns (PAMPs) including lipopolysaccharide (LPS), peptidoglycan and flagellin which stimulate immune responses especially through pattern recognition receptors such as toll-like receptors (TLRs) (Lovewell et al., 2014).

3. Role of Biomolecules on *Pseudomonas aeruginosa*

A complex based pharmacophore model as scaffold could be used as a potential selective inhibitor of uridine diphosphate N-acetylglucosamine (UDP-GlcNAc) acetyltransferase (PaLpxA) which is an essential enzyme produced by the *Pseudomonas* (Bhaskar et al., 2020). Lipopeptides like octapeptin A3, polymixin and colistin binding to component sputum biomolecules may potentially reduce the cell numbers in this bacteria's biofilm (Futschik et al., 2018). The utility of photodynamic therapy using cationic porphyrin on the treatment of bacterial infections by *P. aeruginosa* resulted in the damage of biofilms by direct killing of individual cells (Collins et al., 2010). The invasion and cytotoxicity of this pathogen are mutually exclusive events involving protein tyrosine kinase (PTK) activity and its inhibitor may have a potential therapeutic ability to block their pathways on the host cell (Evans et al., 1998).

4. Biomolecules as Immunomodulators

Autophagy serves as an effector and regulator in macrophages during immune response against pathogen invasion for the elimination of invasive bacteria through autolysosome degradation leading to pathogen clearance. Drugs such as rapamycin and IFN-γ are autophagy inducers which are found to be augmenting the bacterial clearance by autophagy (Yuan et al., 2012). Bacterial growth and lethality of *P. aeruginosa* was increased by X-irradiation but not by carrageenan treatment and it was phagocytized and killed efficiently by polymorphonuclear cells and macrophages (Tasukawa et al., 1979). Exotoxin-A of *P. aeruginosa* was toxic to human peripheral blood macrophages *in vitro*, however this cytotoxicity has been found to be neutralizable with anti-exotoxin serum (Pollack et al., 1978). IgG antibodies seemed to be the principle opsonin in normal respiratory secretions with the specific agglutinative activity that facilitated the phagocytosis and killing of this bacteria by alveolar macrophages (Reynolds et al., 1973). MUC1 is membrane-tethered mucin that interacts with these bacteria through flagellin and nullifying it in mice cleared this bacterium with the enhanced inflammatory response towards this bacteria and flagellin by resident airway macrophages (Lu et al., 2006). Acid sphingomyelinase which is enriched with ceramide is activated in alveolar macrophages leading to the formation of ceramide-enriched membrane platforms that are required for activation of ROS for phagocytosis (Zhang et al., 2008).

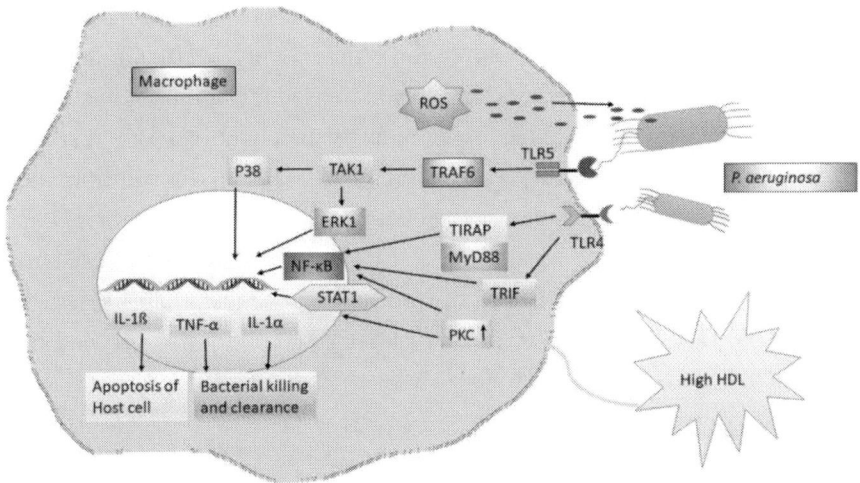

Figure 2. Activation of TLR4 and TLR5 signaling pathways in the host macrophage leading to host cell apoptosis as well as a bacterial clearance and the effect of high-density lipoproteins as a potential immunomodulatory biomolecule capable of exerting pro-inflammatory response on macrophages via PKC-NF-κB/STAT1 signaling.

5. BACTERIAL STRATEGIES FOR PHAGOCYTIC EVASION

The viscous nature of *P. aeruginosa* alginate exerts a non-specific inhibitory effect on the uptake and degradation of phagocytosable particles (Simpson et al., 1988). Exopolysaccharide of mucoid strains of *P. aeruginosa* appears to alter its surface characteristics when compared to non-mucoid strains; thereby rendering them resistance to non-opsonic phagocytosis and providing it a survival advantage (Cabral et al., 1987). Mucoid phenotypes have been indicated as refractile to the ingestion of the pathogens by inhibiting the stimulation of macrophage response to opsonized and unopsonized mucoid strains (Krieg et al., 1988). Previous studies have suggested that flagellar-mediated Toll-like receptor (TLR) recognizes flagellar components like flagellin as phagocytic activation signal and loss of such flagellar expression has been found to enable inactivation of innate immune response thereby causing phagocytic evasion (Balloy et al., 2007). A recent study contradicted this paradigm by demonstrating that this evasion is primarily due to loss of swimming motility and is independent of flagellar expression (Amiel et al., 2010). Gradual loss of bacterial flagellar torsion progressively increased the ability of bacteria to evade phagocytosis since phagocytes react to mechano-sensory response directed by flagellar motility (Lovewell et al., 2011). In addition, two periplasmic superoxide dismutases (SODs) such as SODB (Fe-SOD) and SODM (Mn-SOD) play a significant role in this bacteria's resistance towards NOX-mediated oxidative stress of macrophages. Out of these two enzymes, SODB dominates SODM by playing maximum pro-survival role by modulating macrophage activity (Cavinato et al., 2020).

6. CLINICAL ASPECTS

P. aeruginosa is mainly focused and in high demand for the development of novel therapy due to its remarkable virulent infections against human health. The method of eradication becomes challenging as most strains of *P. aeruginosa* remains resistant to antibiotics and another mode of clinical trials which possess high levels of innate and acquired resistance mechanisms to overcome antibiotics. Also, it has been reported that they acquire some adaptive biofilm mediated resistance forming persistence microbiome (Panga et al., 2019). The excessive use of antibiotics during treatment accelerates the development of multidrug-resistance responsible for recalcitrance and worsening of infections (Hirsch et al., 2010).

Pseudomonas reveals the resistance to various groups of antibiotics including aminoglycosides, quinolones and β-lactams (Hancock et al., 2000). *P. aeruginosa* forms resistance against the antibiotic through increased biofilm formation mediated by bis-(3'-5')-cyclic dimeric guanosine monophosphate (c-di-GMP) biofilm forming messenger (Bouffartigues et al., 2015) and several outer membrane porins, especially the basic amino acid-specific porin OprD, mutation in the nfxB gene is involved in antibiotic uptake (Hancock et al., 2002). During the mutation of nfxB gene or in the absence of this OprD which contains the binding sites for carbapenems a β-lactam subfamily *P. aeruginosa* develops the resistance to this class of antibiotic. In addition, the mutations occurring in a β-lactamase inducible gene ampC, perhaps expressed resistance to cephalosporins (Panga et al., 2019). Some liposomal preparations of levofloxacin inhalation solution which was tested against *in vitro* and animal models propose efficacy against pathogen at the same time has specificity for *P. aeruginosa* (Hurley et al., 2012). Thus, the antibiotic treatment has the disadvantage of multidrug-resistant capability of *P. aeruginosa*, novel strategies are applied against *P. aeruginosa* and they include inhibition of biofilm formation (Panga et al., 2019).

The azithromycin is the only quorum sensing inhibitor that has been tested in clinical trials, *P. aeruginosa* confers resistance to azithromycin through the mutation-caused upregulated MexCD-OprJ expression may reduce the efficacy of azithromycin (Gillis et al., 2005). It is also reported that some bio-lectins can disrupt the initial attachment of *P. aeruginosa* involved in biofilm formation in human lung epithelial cells *in vitro* (Lerrer et al., 2007; Susilowati et al., 2017). These inhibition methods require further assessment in *in vivo* models and clinical trials and also phage therapies as well as electrochemical scaffold methods are proven to be effective against microbiome. However, the clinical trials possess some disadvantages (Sulakvelidze et al., 2001; Vandenheuvel et al., 2015, Sultana et al., 2015).

Considering all these drawbacks involved in the eradication, a new safe and effective therapy need to be employed to overcome multidrug-resistant strains of *Pseudomonas*. The plant based biomolecules like curcumin, resveratrol, epigallocatechol3-gallate, quercetin, colchicine, capsaicin, andrographolide, genistein have potent effects on cellular as well as humoral immune functions in pre-clinical investigations and will highlight their importance in the clinical trial (Ibrahim et al., 2015). Inulin-type fructans (ITFs), possess highly responsive immune activity as well as activation of phagocytosis immunomodulatory responses involved in signaling and antioxidant pathways. These plant derived compounds have documented medicinal health-promoting properties and potential role in immunomodulation (Dobrange et al., 2019). B-Glucans, the polysaccharides isolated from microorganism and plants are known for its immunomodulation property that activates the immune system (Novak et al., 2008).

CONCLUSION AND PERSPECTIVES

P. aeruginosa appears to cause a wide range of multidrug-resistant infections in patients with cystic fibrosis, those who receive immunosuppressive drugs leading to severe lung infections like Ventilator-associated pneumonia (VAP) that accounts for almost 50% of mortality rate. These strains of *P. aeruginosa* forms biofilm and are likely to envelop other bacteria and shows recalcitrant resistance to drugs and therapies. These conditions are life threatening and thus we urge to find effective therapeutics. *P. aeruginosa* are investigated against various new therapeutic approaches viz., the use of bacteriophages, enzyme (phage encoded lytic proteins) and antibiotics. Unfortunately, the mechanism underlying to this resistance is revealed to some extent and at present there is an important need for improvement of preventive and novel therapeutic measures to fight these infections.

The records presented in this article specify that these biomolecules are regulating immunity by enhancing the phagocytic response. This knowledge leads to the effective idea that biomolecules provides potent antimicrobial and also immunomodulatory activities of macrophages which acts as a major strategy against microbes. As already confirmed, the biomolecules are involved in producing non-host mediated antibacterial immune response providing evidence of agglutination and opsonisation mechanism for defence. In addition, there is also sufficient indication that suggests functionally active molecules may up regulate activation of immune responses by influencing the immune system through bactericidal effector functions such as signaling, stimulation, modulation of macrophages, lymphocytes, and cytokine production. Further, some clinical reports also show that these biomolecules like flavonoids, glycosides, polysaccharides, terpenoids, essential oils, and alkaloids have the efficacy against infection. The interaction of biomolecules in *in vitro* and animal models with clinically relevant doses has a potential effect as well as safe and non-resistant. Finally, this may also lead to the development of novel therapies on biomolecule based immunomodulation in the future.

SUMMARY

This review will spotlight the current knowledge in phagocytic recognition and clearance of bacteria in the context of phagocytic receptors as well as pathways by immunomodulatory action of biomolecules. The bacterial strategies for phagocytic avoidance and inflammatory mechanisms altering phagocytic clearance of bacteria will be highlighted. This knowledge leads to the effective idea that biomolecules provide the potential antimicrobial and also immunomodulatory activities of macrophages which acts as a major strategy against microbes. As already confirmed, the biomolecules are involved in producing non-host mediated antibacterial immune response providing the evidence of agglutination and opsonisation mechanism for defence. In addition, there is also sufficient indication suggesting that those functionally active molecules may up regulate activation of immune responses by influencing the immune system through bactericidal effector functions such as signaling, stimulation, and modulation of macrophages, lymphocytes, and cytokine production. Further, some clinical reports also show that these biomolecules like flavonoids, glycosides, polysaccharides, terpenoids, essential oils, and alkaloids have the efficacy against the infection in *in vitro* and animal models with clinically relevant doses that have the potential effect, are safe and non-resistant. Finally, this may also lead to the development of novel therapies on biomolecule based immunomodulation in the near future.

There is little doubt regarding the potential of biomolecules in the development of new food products, nutraceuticals, and diets. Studies regarding the biochemistry of these molecules show that the extraction of selected compounds can greatly benefit human health. However, while our understanding of these compounds keeps expanding, it is important to monitor how they can be introduced in the market and our diets, being sure that no nefarious effects can manifest. Consumption of plant-based supplements, under proper guidance, keeping in mind its interactions, might help us prevent, and possibly cure many diseases. Further studies and clinical trials focusing on its side effects, interactions with other supplements, and long-term toxicity, if proved safe, may make possible the prevention and eradication of many life-threatening diseases.

REFERENCES

Amiel, Eyal, Rustin R. Lovewell, George A. O'Toole, Deborah A. Hogan, and B. Berwin. 2010. "*Pseudomonas Aeruginosa* Evasion of Phagocytosis Is Mediated by Loss of Swimming Motility and Is Independent of Flagellum Expression." *Infection and Immunity* 78(7). doi: 10.1128/IAI.00144-10.

Ballinger, Megan N., Robert Paine III, Carlos H. C. Serezani, David M. Aronoff, Esther S. Choi, Theodore J. Standiford, Galen B. Toews, and Bethany B. Moore. 2006. "Role of Granulocyte Macrophage Colony-Stimulating Factor during Gram-Negative Lung Infection with Pseudomonas aeruginosa." *American Journal of Respiratory Cell and Molecular Biology* 34(6). doi: 10.1165/rcmb.2005-0246OC.

Balloy, Viviane, Amrisha Verma, Sudha Kuravi, Mustapha Si-Tahar, Michel Chignard, and Reuben Ramphal. 2007. "The Role of Flagellin versus Motility in Acute Lung Disease Caused by *Pseudomonas aeruginosa*." *The Journal of Infectious Diseases* 196(2). doi: 10.1086/518610.

Baskar, Baki Vijaya, Tirumalasetty Muni Chandra Babu, Aluru Rammohan, Gui Yu Zheng, Grigory V. Zyryanov, and Wei Gu. 2020. "Structure-Based Virtual Screening of *Pseudomonas aeruginosa* LpxA Inhibitors Using Pharmacophore-Based Approach." *Biomolecules* 10(2). doi: 10.3390/biom10020266.

Bassetti, Matteo, Antonio Vena, Antony Croxatto, Elda Righi, and Benoit Guery. 2018. "How to manage *Pseudomonas aeruginosa* infection." *Drugs in Context* 7. doi: 10.7573/dic.212527.

Bodey, Gerald P., Ricardo Bolivard, Victor Fainstein, and Leena Jadeja. 1983. "Infections Caused by *Pseudomonas aeruginosa*." *Reviews of Infectious Diseases* 5(2). doi: 10.1093/clinids/5.2.279.

Bouffartigues, Emeline, Joana A. Moscoso, Rachel Duchesne, Thibaut Rosay, Laurene Fito-Boncompte, Gwendolkine Gicquel, Olivier Maillot, Magalie Bernard, Alexis Bazire, Gerald Brenner-Weiss, Olivier Lesouhaitier, Patrice Lerouge, Alain Dufour, Nicole Orange, Marc G. J. Feuilloley, Joery Overhage, Alain Filoux, and Sylvie Chevalier. 2015. "The Absence of the *Pseudomonas aeruginosa* OprF Protein leads to Increased Biofilm Formation through Variation in c-di-GMP Level." *Frontiers in Microbiology* 6. doi: 10.3389/fmicb.2015.00630.

Cabral, David A., Bernadette A. Loh, and David P. Speert. 1987. "Mucoid *Pseudomonas aeruginosa* Resists Nonopsonic Phagocytosis by Human Neutrophils and Macrophages." *Pediatric Research* 22(4). doi: 10.1203/00006450-198710000-00013.

Cavinato, Luca, Elena Genise, Francesco R. Luly, Enea G. Di Domenico, Paola Del Porto, and Fiorentina Ascenzioni. 2020. "Escaping the Phagocytic Oxidative Burst: the Role of SODB in the Survival of *Pseudomonas aeruginosa* Within Macrophages." *Frontiers in Microbiology* 11(326). doi: 10.3389/fmicb.2020.00326.

Cheung, Dororthy O. Y., Keith Halsey, and David Paul Speert. 2000. "Role of Pulmonary Alveolar Macrophages in Defense of the Lung against *Pseudomonas aeruginosa*." *Infection and Immunity* 68. doi: 10.1128/IAI.68.8.4585-4592.2000.

Chouhan, Garima, Mohammad Islamuddin, Dinkar Sahal, and Farhat Afrin. 2014. "Exploring the Role of Medicinal Plant-based Immunomodulators for Effective Therapy of Leishmaniasis." *Frontiers in Immunology* 5(193). doi: 10.3389/fimmu.2014.00193.

Collins, Tracy L., Elizabeth A. Markus, Daniel J. Hassett, and Jayne B. Robinson. 2010. "The Effect of a Cationic Porphyrin on *Pseudomonas aeruginosa* Biofilms." *International Journal of Current Microbiology and Applied Sciences* 61(5). doi: 10.1007/s00284-010-9629-y.

Dobrange, Erin, Darin Peshev, Bianke Loedolff, and Wim Van den Ende. 2019. "Fructans as Immunomodulatory and Antiviral Agents: The Case of *Echinacea*" Biomolecule 9(10). doi: 10.3390/biom9100615.

Driscoll, James A., Steven L. Brody, and Marin H. Kollef. 2007. "The Epidemiology, Pathogenesis and Treatment of *Pseudomonas aeruginosa* Infections." *Drugs* 67. doi: 10.2165/00003495-200767030-00003.

Elhelu, Mohamed A. 1983. "The Role of Macrophages in Immunology." *Journal of National Medical Association* 75 (3). PMID: 6343621.

Engels, W., J. Endert, Miriam Kamps, and C. P. A. Van Boven. 1985. "Role of Lipopolysaccharide in Opsonization and Phagocytosis of *Pseudomonas aeruginosa.*" *Infection and Immunity* 49(1). doi: 10.1128/IAI.49.1.182-189.1985.

Evans, David J., Dara W. Frank, Viviane Finck-Barbancon, Christine Wu, and Suzzane M. J. Fleiszig. 1998. "*Pseudomonas aeruginosa* Invasion and Cytotoxicity are Independent Events, Both of which Involve Protein Tyrosine Kinase Activity." *Infection and Immunity* 66(4). doi: 10.1128/IAI.66.4.1453-1459.1998.

Finlay, Brett B., and Stanley Falkow. 1988. "Comparison of the Invasion Strategies Used by *Salmonella Cholerae-Suis, Shigella Flexneri* and *Yersinia Enterocolitica* to Enter Cultured Animal Cells: Endosome Acidification Is Not Required for Bacterial Invasion or Intracellular Replication." *Biochimie* 70. doi: 10.1016/0300-9084(88)90271-4.

Fleiszig, S. M. J., T. S. Zaidi, and G. B. Pier. 1995. "*Pseudomonas aeruginosa* Invasion of and Multiplication within Corneal Epithelial Cells *In Vitro.*" *Infection and Immunity* 63. PMID: 7558321.

Gillis, Richard J., Kimberly G. White, Kyoung-Hee Choi, Victoria E. Wagner, Herbert P. Schweizer, and Barbara H. Iglewski. 2005. "Molecular Basis of Azithromycin-Resistant *Pseudomonas aeruginosa* Biofilms." *Antimicrobial Agents and Chemotherpay* 49. doi: 10.1128/AAC.49.9.3858-3867.2005.

Hancock, Robert E. W., and David P. Speert. 2000. "Antibiotic Resistance in *Pseudomonas aeruginosa*: Mechanisms and Impact on Treatment." *Drug Resistance Updates* 3. doi: 10.1054/drup.2000.0152.

Hancock, Robert E. W., and Fiona S. L. Brinkman. 2002. "Function of *Pseudomonas* Porins in Uptake and Efflux." *Frontiers in Microbiology* 56. doi: 10.1146/annurev.micro.56.012302.160310

Hazlett, Linda D., Sharon A. McClellan, Ronald P. Barrett, Xi Huang, Yunfan Zhang, Minhao Wu, Nico van Rooijen, and Elizabeth Szliter. 2010. "IL-33 Shifts Macrophage Polarization, Promoting Resistance against *Pseudomonas aeruginosa* Keratitis." *Investigative Ophthalmology & Visual Science* 51. doi: 10.1167/iovs.09-3983.

Hirsch, Elizabeth B., and Vincent H. Tam. 2010. "Impact of Multidrug-resistant *Pseudomonas aeruginosa* Infection on Patient Outcomes." *Expert Review of Pharmacoeconomics & Outcomes Research* 10. doi: 10.1586/erp.10.49.

Hurley, Matthew N., Miguel C. Mara, and Alan R. Smyth. 2012. "Novel Approaches to the Treatment of *Pseudomonas aeruginosa* Infections in Cystic Fibrosis." *The European Respiratory Journal* 40. doi: 10.1183/09031936.00042012.

Jantan, Ibrahim, Waqas Ahmed, and Syed Nasir Abbas Bukhari. 2015. "Plant-derived Immunomodulators: an Insight on their Preclinical Evaluation and Clinical Trials." *Frontiers in Plant Science* 25 (6). doi: 10.3389/fpls.2015.00655.

Kernacki, Karen A., Ronald P. Barett, Jeffery A. Hobden, and Linda D. Hazlett. 2000. "Macrophage Inflammatory Protein-2 is a Mediator of Polymorphonuclear Neutrophil Influx in Ocular Bacterial Infection." *The Journal of Immunology* 164 (2). doi: 10.4049/jimmunol.164.2.1037.

Krieg, D. P., R. J. Hemke, Victor F. German, and J. A. Margos. 1988. "Resistance to Mucoid *Pseudomonas aeruginosa* to Non-opsonic

Phagocytosis by Alveolar Macrophages *In Vitro*." *Infection and Immunity* 56(12). doi: 10.1128/IAI.56.12.3173-3179.1988.

Kurahashi, Kiyoyasu, Osamu Kajikawa, Teiji Sawa, Maria Ohara, Michael A. Gropper, Dara W. Frank, Thomas R. Martin, and Jeanine P. Weiner-Kronish. 1999. "Pathogenesis of Septic Shock in *Pseudomonas aeruginosa* Pneumonia." *The Journal Clinical Investigation* 104(6). doi: 10.1172/JCI7124.

Lerrer, Batia, Keren D. Zinger-Yosovich, Benjamin Avrahami, and Nechama Gilboa-Garber. 2007. "Honey and Royal Jelly, like Human milk, Abrogate Lectin-dependent Infection-preceding *Pseudomonas aeruginosa* Adhesion." *The ISME Journal* 1. doi: 10.1038/ismej.2007.20.

Li, Dawei, Bing Zhou, and Bei Lv. 2020. "Antibacterial Therapeutic Agents composed of Functional Biological Molecules." *Journal of Chemical Education* 2020. doi: 10.1155/2020/6578579.

Lima, Flavia Luna, Paulo Pinto Joazeiro, Marcelo Lanelloti, Luciana Maria Hollanda, Bruna de Araujo Lima, Edlaine Linares, Ohara Augusto, Marcelo Brocchia, and Selma Giorgio. 2015. "Effects of Hyperbaric Oxygen on *Pseudomonas aeruginosa* Susceptibility to Imipenem and Macrophages." *Future Microbiology* 10(2). doi: 10.2217/fmb.14.111.

Lovewell, Rustin R., Ryan M. Collins, Julie L. Acker, George A. O'Toole, Matthew J. Wargo, and Brent Berwin. 2011. "Step-wise Loss of Bacterial Flagellar Torsion Confers Progressive Phagocytic Evasion." *PLOS Pathogens* 7(9). doi: 10.1371/journal.ppat.1002253.

Lovewell, Rustin R., Yash R. Pattankar, and Brent Berwin. 2014. "Mechanisms of Phagocytosis and Host Clearance of *Pseudomonas aeruginosa*." *American Journal of Physiology-Lung Cellular and Molecular Physiology* 306(7). doi: 10.1152/ajplung.00335.2013.

Lu, Wenju, Akinori Hisatsune, Takeshi Koga, Kosuke Kato, Ippei Kuwahara, Erik P. Lillehoj, Wilbur H. Chen, Alan S. Cross Sandra J. Gendler, Andrew T. Gewirtz, and Kwang Chul Kim. 2006. "Cutting Edge: Enhanced Pulmonary Clearance of *Pseudomonas aeruginosa* by muc1 Knockout Mice." *Journal of Immunology Research* 176. doi: 10.4049/jimmunol.176.7.3890.

Meshulam, T., H. A. Yerbrugh, and J. Verhoef. 1982. "Opsonisation and Phagocytosis of Mucoid and Non-mucoid *Pseudomonas aeruginosa* Strains." *European Journal of Clinical Microbiology & Infectious Diseases* 1. doi: 10.1007/BF02014202.

Mukopadhyay, Manas Kumar, Pratyusha Banerjee, and Debjani Nath. 2012. "Phytochemicals-biomolecules for the Prevention and Treatment of Human Diseases a Review." *International Journal of Scientific & Engineering Research* 3(7).

Nieuwenhuis, Edaward E. S., Tetsuya Matsumoto, Mark Exley, Robbert A. Schleipman, Jonathan Glickman, Dan T. Bailey, Nadia Corazza, Sean P. Colgan, Andrew B. Onderdonka, and Richard S. Blumberg. 2002. "CD1d-dependent Macrophage-mediated Clearance of *Pseudomonas aeruginosa* from Lung." *Nature Medicine* 8. doi: 10.1038/nm0602-588.

Novak. M., and Vaclav Vetvicka. 2008. "β-Glucans, History, and the Present: Immunomodulatory Aspects and Mechanisms of Action." *Journal of Immunotoxicology* 5. doi: 10.1080/15476910802019045.

Pang, Zheng, Renee Raudonisb, Bernard R. Glickc, Tong-Jun Lina, and Zhenyu Chenga. 2019. "Antibiotic Resistance in *Pseudomonas aeruginosa*: Mechanisms and Alternative Therapeutic Strategies." *Biotechnology Advances* 37. doi: 10.1016/j.biotechadv.2018.11.013.

Pollack, M., and S. E. Anderson Jr. 1978. "Toxicity of *Pseudomonas aeruginosa* exotoxin A for Human Macrophages." *Infection and Immunity* 19(3). PMID: 417028.

Restrepo, Clara I., Qun Dong, Jordon Savov, William I. Mariencheck, and Jo Rae Wright. 1999. "Surfactant Protein D Stimulates Phagocytosis of *Pseudomonas aeruginosa* by Alveolar Macrophages." *American Journal of Respiratory Cell and Molecular Biology* 21. doi: 10.1165/ajrcmb.21.5.3334.

Reynolds, Herbert. Y., John A. Kazmierowski, and Harold H. Newball. 1975. "Specificity of Opsonic Antibodies to Enhance Phagocytosis of *Pseudomonas aeruginosa* by Human Alveolar Macrophages." *The Journal of Clinical Investigation* 56. doi: 10.1172/JCI108102.

Reynolds, Herbert Y., and Russell E. Thompson. 1973. "Pulmonary Host Defences: II. Interaction of Respiratory Antibodies with *Pseudomonas aeruginosa* and Alveolar Macrophages." *Journal of Immunology Research* 111(2). PMID: 4197983.

Rizza, Marco D., Paola D. Della, Rafael Narancio, Andrea Cabrera, and Fernando Ferreira. 2012. "Biomolecules as Most Defence Weapon against Microbial Pathogens." *Recent patents on DNA & gene sequences* 2(2). doi: 10.2174/187221508784534186.

Rogers, Kara. 2020. Biomolecule. *Britannica*.

Rudner, Xiaowen L., Karen A. Kernacki, Ronald P. Barrett, and Linda D. Hazlett. 2000. "Prolonged Elevation of IL-1 in *Pseudomonas aeruginosa* Ocular Infection Regulates Macrophage-Inflammatory Protein-2 Production, Polymorphonuclear Neutrophil Persistence, and Corneal Perforation." *Journal of Immunology* 164 (12). doi: 10.4049/jimmunol.164.12.6576.

Schneider-Futschik, Elena K., Olivia K. A. Paulin, Daniel Hoyer, Kade D. Roberts, James Ziogas, Mark A. Baker, John Karas, Jian Li, and Tony Velkov. 2018. "Sputum Active Polymyxin Lipopeptides: Activity against Cystic Fibrosis *Pseudomonas aeruginosa* Isolates and Their Interactions with Sputum Biomolecules." *ACS Infectious Diseases* 4(5). doi: 10.1021/acsinfecdis.7b00238.

Simon, Daniel, and Richard F. Rest. 1992. "Escherichia coli Expressing a *Neisseria gonorrhoea* Opacity-associated Outer Membrane Protein Invade Human Cervical and Endometrial Epithelial Cell Lines." *Proceedings of the National Academy of Sciences* 89. doi: 10.1073/pnas.89.12.5512.

Simpson, Jeremy A., Susan C. Smith, and Roger T. Dean. 1988. "Alginate Inhibition of the Uptake of *Pseudomonas aeruginosa* by Macrophages." *Microbiology* 134(1). doi: 10.1099/00221287-134-1-29.

Speert, David P., Samuel D. Wright, Samuel C. Silverstein, and Bernadette Mah. 1988. "Functional Characterization of Macrophage Receptors for In Vitro Phagocytosis of Unopsonized *Pseudomonas aeruginosa*." *Journal of Clinical Investigation* 82(3). doi: 10.1172/JCI113692.

Sulakvelidze, Alexander, Zemphira Alavidze, and J. Glenn Morris Jr. 2001. "Bacteriophage Therapy." *Antimicrobial Agents and Chemotherapy* 45. doi: 10.1128/AAC.45.3.649-659.2001.

Sultana, Sujala T., Eehan Atci, Jerome T. Babauta, Azeza Mohamed Falghoush, Kevin R. Snekvik, Douglas R. Call, and Haluk Beyenal. 2015. "Electrochemical Scaffold Generates Localized, Low Concentration of Hydrogen Peroxide that Inhibits Bacterial Pathogens and Biofilms." *Scientific Reports* 5. doi: 10.1038/srep14908.

Susilowati, Heni, Keiji Murakami, Hiromichi Yumoto, Takashi Amoh, Kouji Hirao, Katsuhiko Hirota, Takashi Matsuo, and Yoichiro Miyake. 2017. "Royal Jelly Inhibits *Pseudomonas aeruginosa* Adherence and Reduces Excessive Inflammatory Responses in Human Epithelial Cells." *BioMed Research International* 2017(3191752). doi: 10.1155/2017/3191752.

Tatsukawa, Keiji, Masao Mitsuyama, Kenji Takeya, and Kikuo Nomoto. 1979. "Differing Contribution of Polymorphonuclear Cells and Macrophages to Protection of Mice against Listeria monocytogenes and *Pseudomonas aeruginosa*." *Microbiology* 115(1). doi: 10.1099/00221287-115-1-161.

Vandenheuvel, Dieter, Rob Lavigne, and Harald Brussow. 2015. "Bacteriophage Therapy: Advances in Formulation Strategies and Human Clinical trials." *Annual Review of Virology* 2. doi: 10.1146/annurev-virology-100114-054915.

Yuan, Kefei, Canhua Huang, John Fox, Donna Laturnus, Edward Carlson, Binjie Zhang, Qi Yin, Hongwei Gao, and Min Wug. 2012. "Autophagy Plays an Essential Role in the Clearance of *Pseudomonas aeruginosa* by Alveolar Macrophages." *Journal of Cell Science* 125. doi: 10.1242/jcs.094573.

Zhang, Yang, Xiang Li, Alexander Carpinteiro, and Ereich Gulbins. 2008. "Acid Sphingomyelinase Amplifies Redox Signaling in *Pseudomonas aeruginosa*-induced Macrophage Apoptosis." *Journal of Immunology Research* 181(6). doi: 10.4049/jimmunol.181.6.4247.

Zhou, Zimei, Ronald P. Barrett, Sharon A. McClellan, Yunfan Zhang, Elizabeth A. Szliter, Nico van Rooijen, and Linda D. Hazlett. 2008. "Substance P Delays Apoptosis, Enhancing Keratitis after *Pseudomonas aeruginosa* Infection." *Investigative Ophthalmology &Visual Science* 49. doi: 10.1167/iovs.08-1906.

Chapter 5

SURVIVAL STRATEGIES OF *PSEUDOMONAS AERUGINOSA* IN OCULAR INFECTIONS

Vidyarani Mohankumar[1,*,†], *Kathirvel Kandasamy*[1], *Prajna Lalitha*[2] *and Bharanidharan Devarajan*[1,*,‡]

[1]Department of Microbiology and Bioinformatics,
Aravind Medical Research Foundation, Tamil Nadu, India
[2]Department of Clinical Microbiology, Aravind Eye Hospital,
Tamil Nadu, India

ABSTRACT

Ocular infections caused by *Pseudomonas aeruginosa* progress rapidly and result in worse clinical outcomes. The bacterial exotoxins, proteases, and other virulence factors cause host cell death, progressive tissue damage and induction of an inflammatory response. The high virulence of the organism is complemented by an array of innate and

---

* Both have equally contributed.
† Corresponding Author's Email: mvidhya@gmail.com.
‡ Corresponding Author's Email: bharani@aravind.org.

acquired drug resistance mechanisms. These include target site alteration, enzymatic drug inactivation, alterations in membrane permeability, drug efflux as well as biofilm formation. In addition, drug tolerance prevents complete bacterial clearance from the tissues despite effective antibiotic treatment. Most drug-sensitive isolates of *P. aeruginosa* form persisters that survive and replicate after antibiotic treatment.

Further, *P. aeruginosa* invades host cells to escape immune recognition and subsequent degradation. Studies on intracellular *P. aeruginosa* survival in human corneal epithelial cells implicate autophagy as an important innate defense mechanism. Experimental manipulation of the autophagic process resulted in variable intracellular bacterial load, which may have therapeutic value in *P. aeruginosa* corneal ulcers. Understanding the intracellular bacterial survival and drug tolerance mechanisms may help to select appropriate therapy for *P. aeruginosa* infections.

Keywords: ocular infections, Pseudomonas, survial statergies, antibiotic resistance

1. INTRODUCTION

Pseudomonas aeruginosa, a Gram-negative opportunistic pathogen causes acute and chronic infections, especially in immunocompromised patients. Among Gram-negative bacteria, *P. aeruginosa* is the predominant one causing ocular infections (Lalitha et al., 2014). *P. aeruginosa* keratitis is an ocular inflammatory disease that may lead to the severe visual disability with corneal scarring (Keay et al., 2006). Corneal ulcers caused by *P. aeruginosa* have been described as more severe than the other bacterial corneal ulcers. Moreover, Pseudomonas keratitis is difficult to treat and results in poor visual outcome compared to other bacterial corneal ulcers (Krachmer et al., 1997; Green et al., 2008; Sy et al., 2012). In South India, *P. aeruginosa* is responsible for 8 to 21% of bacterial keratitis cases (Sy et al., 2012). Risk factors for bacterial keratitis are those that disrupt the integrity of the corneal epithelium which includes contact lens wear, ocular trauma, previous ocular surgery, surgical sutures and ocular surface diseases. If the therapy is not promptly initiated, it leads to vision loss (Keay et al., 2006). Virulence, antibiotic resistance, multi-drug tolerance and evasion of host defense responses are among the major traits contributing to the pathogenic potential of *P. aeruginosa*.

2. VIRULENCE OF *P. AEUROGINOSA*

Pseudomonas aeruginosa has a range of virulence determinants that contribute to its pathogenicity. Bacterial adhesion to the cells is mediated by structures like flagella, pili, fimbriae and lipopolysaccharide (LPS) (Zolfaghar et al., 2003). This is followed by the secretion of bacterial exo proteins or proteases that cleave host cell surface molecules and mediate host cell death, and or induction of an inflammatory response (Marquart et al., 2013). The *P. aeruginosa* type III secretion system (T3SS) plays an important role during the infection of eukaryotic host cells. The T3SS forms complex needle-like machines on the bacterial surface that function in a highly regulated manner to transport proteins and toxins into host cells (Hauser et al., 2009). The four T3SS toxins are exotoxin S (Exo S), Exo T, Exo U and Exo Y (Finck-Barbançon et al., 1997; Fleiszig et al., 1997; Yahr et al., 1998). Type III secretion system was identified as a significant virulence mechanism in the pathogenesis of Pseudomonal keratitis (Fleiszig et al., 1995; Finck-Barbançon et al., 1997; Shen et al., 2015). A *P. aeruginosa* isolate can be categorized as an invasive or cytotoxic type based on the type of T3SS exotoxin secreted since they affect corneal epithelial cells differently. Invasive strains that carry the exoS gene invade corneal epithelial cells, while cytotoxic strains cause rapid host cell necrosis with the help of exoU phospholipase (Fleiszig et al., 1996; Fleiszig et al., 1997; Shen et al., 2015). The presence of T3SS toxin-encoding genes in clinical isolates from different infections is associated with differences in bacterial virulence (Feltman et al., 2001) and clinical outcomes (Hauser et al., 2002). The comparative genomics of ocular isolates are shown in Figure 1. The T3SS exotoxin ExoU was detected in three of six *P. aeruginosa* keratitis isolates (BK1 to BK6) sequenced at Aravind Medical Research Foundation, of which two (BK4 and BK5) were reported with poor clinical outcome (underwent Therapeutic Penetrating Keratoplasty) (Kathirvel et al., 2020). In a study by Subedi et al., (2018), seven of eight Indian *P. aeruginosa* keratitis isolates (PA31, PA32, PA33, PA35, PA37 and PA82) were reported to have ExoU, but it was not detected among Australian isolates (PA17, PA40, PA149 and PA175) (Figure 1). Murugan et al., (2016) also reported the presence of ExoU in the MDR isolate VRFPA04, which was isolated from an Indian keratitis patient with poor outcome. In *P. aeruginosa* ocular infections, tissue damage is mediated by cytotoxins like ExoU and ExoS (Fleiszig et al., 1997; Feltman et al., 2001), as well as enzymes like elastase B (Lau et al., 2005; Sadikot et al., 2005), protease IV (Hobden, 2002), *P. aeruginosa* small protease (Thibodeaux et al., 2007) and alkaline protease (Goodman et al., 2004; Tingpej et al., 2007). Proteases contribute to pathogenesis in keratitis through the destruction of connective tissue and the degradation of host immunological factors (Engel et al., 1998).

In addition, the type IV pilus associated genes PilB and PilC were identified in all reported keratitis isolates with partial sequence similarity against PAO1 (Figure 1). The pyoverdine bio-synthetic genes PvdD and PvdE reported (Suzuki T et al., 2018) to have a role in proliferation and invasion on ocular surfaces was found in the majority of reported keratitis isolates with less sequence similarity, except BK1, BK4, PA17 and PA82 which contains the complete sequence (Figure 1). Notably, BK1 and BK4 isolates were collected from keratitis patients with poor clinical outcome. The B-band LPS synthesis O-antigen genes (wzx, wbpA, wbpB, wbpD, wbpE, wbpG, wbpH, wbpI, wbpJ, wbpK) were found to be complete loss in all keratitis isolates except the MDR strain VRFPA04. The flagellar genes, fliC and fliD reported to induce inflammasome and impair bacterial clearance was identified only in Indian isolates BK4 and BK5 which were collected from keratitis patients with poor clinical outcome and also in Australian keratitis isolates PA149 and PA175.

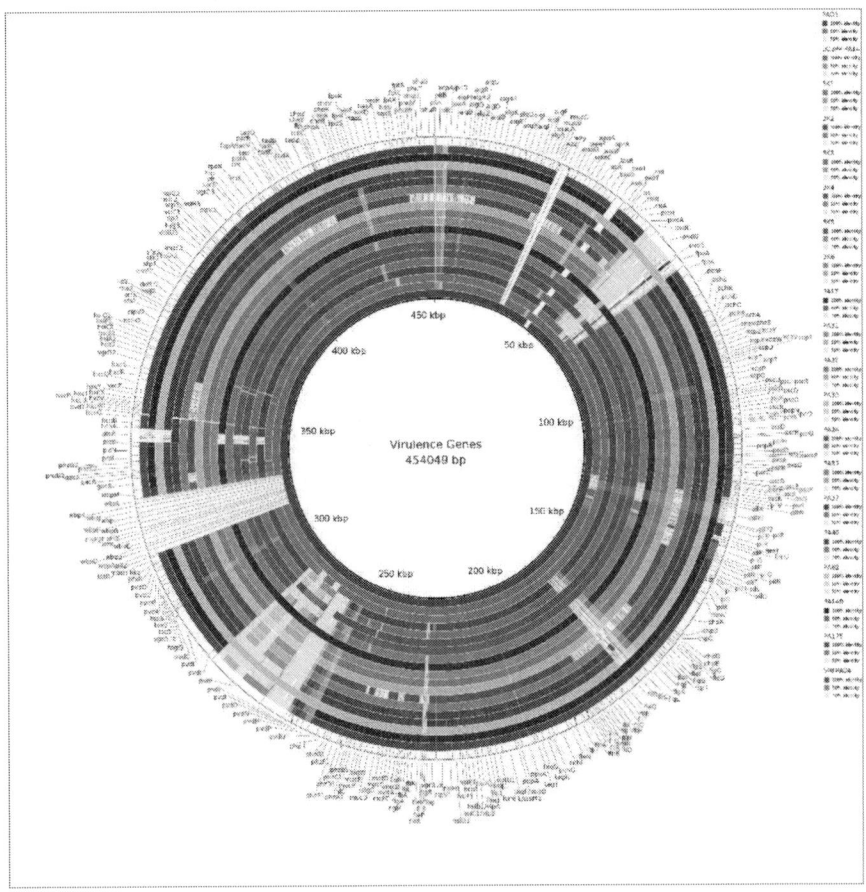

Figure 1.

3. ANTIBIOTIC RESISTANCE

P. aeruginosa develops multidrug resistance to a broad spectrum of antibiotics that poses a major challenge for clinicians. Globally, the inappropriate usage of antibiotics and prolonged treatment duration leads to the development of drug resistance among ocular pathogens (Sharma et al., 2011). In addition to the chromosomally encoded resistance mechanisms, *P. aeruginosa* can acquire resistance genes from other bacteria (Subedi et al., 2017). Apart from these the bacterium also causes target site alteration, enzymatic drug inactivation, alterations in membrane permeability, drug efflux as well as biofilm formation (Poole et al., 2011). Among the different T3SS genotypes, the cytotoxic strains possessing the *exoU* gene were generally found to be more multidrug-resistant in ocular infections (Borkar et al., 2014; Lakshmi Priya et al., 2015). The Steroids for Corneal Ulcer Trial (SCUT) done in bacterial keratitis patients has shown elevated fluoroquinolone resistance in cytotoxic *P. aeruginosa* strains that were associated with worse clinical outcomes (Borkar et al., 2014).

To survive during hostile conditions, *P. aeruginosa* has developed selective outer membrane permeability and an array of efflux systems that maintain cell homeostasis. Among the different efflux systems in *P. aeruginosa*, the resistance-nodulation-division (RND) efflux system has broad substrate specificity with clinical relevance. It mediates active efflux of various antibiotics like β-lactams, chloramphenicol, trimethoprim, fluoroquinolones, tetracyclines and aminoglycosides (Masuda et al., 2000). Target site mutations along with overexpression of RND efflux pumps are associated with multi-drug resistance in *P. aeruginosa* (Shigemura et al., 2015).

Fluoroquinolones primarily target bacterial enzymes like DNA gyrase and topoisomerase IV that are involved in DNA replication, transcription and recombination. ParC, a subunit of topoisomerase IV, shares high sequence similarities in the quinolone resistance determining regions with the gyrA gene. While mutations in DNA gyrase lead to quinolone resistance, the parC mutants increase resistance levels by acting as a prerequisite for gyrA mutations (Muñoz et al., 1996). The predominant amino acid substitutions noted among ocular MDR isolates were Thr-83→Ile in GyrA and Ser-87→Leu in ParC (Thirumalmuthu et al., 2019). The same mutations have previously been reported in ciprofloxacin resistant *P. aeruginosa* isolates and an association was found between ciprofloxacin MIC and the number of target site alterations (Lee ta al., 2005; Nouri et al., 2016). In Gram negative bacteria, mutations first arise in the most susceptible DNA gyrase gene, followed by additional mutations that further increase resistance (Jacoby et al., 2005).

Alterations such as methylation in the 16S rRNA subunit confer pan aminoglycoside resistance in *P. aeruginosa* (Yu et al., 2010). In addition, enzymes that catalyze modification of aminoglycosides are more prevalent among Gram-negative bacteria. Aminoglycoside resistance among the ocular *P. aeruginosa* isolates is largely mediated through the enzymatic modification of drugs (Thirumalmuthu et al., 2019). AME could be acetyltransferases (AACs), nucleotidyltranferases (ANTs) or phosphotransferases (APHs). These enzymes modify the hydroxyl or amino groups of the 2-deoxystreptamine nucleus or the sugar moieties (Ramirez et al., 2010). Among ocular multi drug-resistant isolates, aph(3")-I and aph(6)-I genes were predominant and the presence of AME genes was associated with widespread resistance to gentamicin, amikacin and tobramycin.

Another mechanism by which *P. aeruginosa* resists antibiotics is by forming biofilms. Biofilms are surface-attached bacterial communities that show increased resistance to antimicrobials which could be a thousandfold more than in free floating planktonic cells (Mah et al., 2003). These surface-associated populations are enclosed by the self-secreted extracellular polysaccharide matrix, protein and nucleic acids (Flemming et al., 2010; Balasubramanian et al., 2012). By creating a physical barrier to antimicrobials, biofilms could facilitate the development of drug resistance by decreasing the clinical efficiency of antibiotics. Ocular *P. aeruginosa* have been shown to form biofilms of varying densities, but a direct association with antibiotic resistance was not ascertained (Lakshmi Priya et al., 2015; Thirumalmuthu et al., 2019).

3.1. Drug Tolerance

P. aeruginosa infection can become difficult to eradicate due to the formation of biofilms. The enhanced antibiotic resistance of biofilms has been attributed to the presence of persister cells in biofilms. Studies have shown that exposure of bacteria to intermittent concentrations of antibiotics and the formation of persisters often precedes the evolution of drug resistance (Levin-Reisman et al., 2017). Persisters are a small population of dormant non-growing bacteria that survive antibiotics and also revert to susceptible replicating forms upon antibiotic removal (Bigger et al., 1944; Li et al., 2013; Zhang et al., 2014). By harboring persisters,, biofilms could act as a reservoir of surviving pathogens that may cause relapsing infection (Lewis 2007). Most drug susceptible bacterial populations contain a small fraction of such phenotypic variants that remain viable even after prolonged antibiotic treatments. Since these cells revert to a normal state after reinoculation in antibiotic-free medium, they are not considered as mutants. Moreover, the antibiotic sensitivity of their offsprings remains the same as that of the parent population (Lewis, 2008).

Bacterial persistence was originally discovered by Joseph Bigger, who explored the bacterial killing dynamics of penicillin. When a growing culture of genetically identical bacteria was exposed to a bactericidal antibiotic, it resulted in the rapid killing of a major bulk of the population. But, after a few hours of treatment, there was a dramatic decrease in the killing rate, revealing the existence of a small fraction of persister cells that were phenotypically tolerant to the antibiotic (Bigger, 1944). The descendants of these persister cells had the same antibiotic sensitivity pattern as their ancestors thereby implying bacterial persistence as a non-inherited, epigenetic trait (Bigger, 1944; Keren et al., 2004).

3.2. Drug Tolerance of Persister Cells

Over the years, various hypotheses have been proposed to explain the phenomenon of bacterial persistence. Among them, the most likely explanation lies in the dormant nature of persister cells. Most antibiotics kill bacteria by disrupting the basic cellular processes such as protein translation, transcription, or cell wall synthesis (Lewis, 2008). Unlike drug resistance, the tolerance of persister cells to antibiotics may prevent target corruption by a bactericidal agent by blocking the antibiotic targets. When the bacterial cell-wall synthesis, translation or transcription slows down, the ability of antibiotics in corrupting the function of their target molecules gets compromised. Therefore, at the cost of proliferation, these persister cells can resist killing by the antibiotics. Studies have shown that overproduction of toxins that block cell metabolism increases the persister fraction in a bacterial population (Vázquez-Laslop et al., 2006). Random expression of these toxins due to fluctuations in transcription or translation, could only cause a few cells to become persistent. Such low frequency of persister cells (10^4 to 10^6 of the bacterial population) and the various genes that are associated with the process have delayed in-depth analysis of the phenomenon. Few important genes that are shown to be associated with persistence in *P. aeruginosa* are the global regulators spoT, relA, dksA (Viducic et al., 2006) and rpoS (Murakami et al., 2005). Persisters remain in a quiescent state by downregulating gene expression and complete eradication of these persisters may be possible by enhancing host defense mechanisms along with appropriate antibiotic treatment.

4. INTRACELLULAR SURVIVAL OF *P. AERUGINOSA*

Pseudomonas aeruginosa, primarily considered an extracellular bacterium invades mammalian cells with the help of secretory toxins. Pathogens invade host cells to escape immune recognition and subsequent degradation. The corneal isolates of *P. aeruginosa* including the cytotoxic strains could invade and replicate inside human corneal epithelial cells (Mohankumar et al., 2018). The invading pathogens are in turn captured by either phagosomes or autophagosomes and subsequently delivered to the lysosomes for degradation. T3SS enables intracellular *P. aeruginosa* survival through the formation of membrane blebs that function as a host cell niche for bacterial growth and T3SS effector dependent bacterial survival and replication within perinuclear vacuoles (Angus et al., 2008). Studies using ocular *P. aeruginosa* isolates and T3SS mutants further confirm the role of T3SS toxins in determining the fate of intracellular bacteria in corneal epithelial cells (Mohankumar et al., 2018).

In corneal epithelial cells, the T3SS toxin exoS facilitates escape of intracellular *P. aeruginosa* from acidified vacuoles. Exoenzyme S is an ADP ribosylating protein that has been shown to affect endocytosis and phagosome maturation through Rab5 ADP-ribosylation (Barbieri et al., 2001). The presence of exoS in the host cell cytosol can lead to phagocytosis resistance via actin microfilament disruption (Frithz-Lindsten, 1997). The ADPR activities of exoS and exoT help in subversion of host immune response to promote bacterial survival and progression of corneal disease. Another study by Yan Sun *et al.,* using neutrophils has shown that, the exoS mutants of *P. aeruginosa* PAO1 strain have reduced survival rates *invitro* and an impaired ability to cause neutrophil apoptosis *in vivo*.

4.1. Role of Autophagy in the Intracellular Survival of *P. aeruginosa*

Autophagy, a normal cellular catabolic process involved in protein turnover and degradation of damaged organelles, was shown to play a clear role in the detection and elimination of intracellular pathogens (Mansilla-Pareja et al., 2013). In addition to pathogen clearance, autophagy also regulates inflammasome activation, NFκB activity and interferon production (Levine et al., 2011). On the other hand, LC3-associated phagocytosis (LAP) plays a similar role in pathogen clearance and involves many components of the canonical macroautophagy pathway except the formation of a double membrane autophagosome. As a defense mechanism, the microbes have evolved several ways to antagonize autophagic degradation (Levine, et al., 2011). Occasionally the autophagosomes can also serve as safe niches for the survival of these intracellular bacteria by providing them the necessary nutrients (Deretic and Levine 2009).

P. aeruginosa has been shown to activate autophagy in primary alveolar macrophages and alveolar macrophage cell line through the canonical beclin-1–Atg7–Atg5 pathway (Yuan et al., 2012). The crucial role of autophagy in the immune response against *P. aeruginosa*, and the therapeutic potential of pharmacological agents targeting the autophagy pathway have been demonstrated *in vivo* in *P. aeruginosa* lung infections (Junkins et al., 2013). In ocular infections, autophagy may have a selective role in bacterial clearance owing to the pro-autophagic role of T3SS toxins in corneal epithelial cells. The PAO1Δ*exoST* mutants and T3SS negative clinical strain were less sensitive to the pharmacological modulation of autophagy and had relatively higher replication potential in corneal epithelial cells (Mohankumar et al., 2018). The possibility of an enhanced bacterial survival in the absence of T3SS toxins makes it imperative to reconsider the process of intracellular bacterial clearance in the context of virulence.

Adjunctive therapies manipulating the autophagy pathway may find applications in preventing persistent infections and reducing tissue damage due to inflammation. Various pharmacological agents and antibiotics have been shown to interfere with the autophagy pathway. For example, the macrolide antibiotic azithromycin has been shown to inhibit autophagy in cystic fibrosis patients thereby making them prone to non-tuberculous mycobacterial infections (Renna et al., 2011). The fluoroquinolone Ciprofloxacin has been shown to inhibit radiation induced autophagy and apoptosis in the ileum (Fukumoto et al., 2013). On the other hand, certain antibiotics activate host cell autophagy to clear the intracellular pathogens efficiently (Kim et al., 2012). Based on these observations and relevant literature, it is reasonable to speculate that certain antibiotics may inhibit or activate host defence mechanisms which in turn may decide the fate of the internalized bacteria.

5. *PSEUDOMONAS AERUGINOSA* GENOME

P. aeruginosa is an opportunistic pathogen with an ability to thrive in most natural and man-made environments. The ability to thrive in a broad range of environments is in part caused by the fact that it possesses a large and diverse genome (Stover et al., 2000; Wolfgang et al., 2003). Among all sequenced bacterial genomes, *P. aeruginosa* possesses the largest proportion of regulatory genes (Stover et al., 2000; Wolfgang et al., 2003). This remarkable feature probably provides a means for coordinating the expression of various genes in response to a wide range of environmental demands.

REFERENCES

Angus A. A., Lee A. A., Augustin D. K., et al., 2008. "Pseudomonas aeruginosa Induces Membrane Blebs in Epithelial Cells, Which Are Utilized as a Niche for Intracellular Replication." *Infect Immun.* doi: 10.1128/IAI.01221-07.

Balasubramanian A, Chairman K, Ranjit Singh A. J. A, Alagumuthu G.2012."Isolation and identification of microbes from biofilm of Urinary catheters and antimicrobial Susceptibility evaluation." *Asian Pacific Journal of Tropical Biomedicine.* doi: 10.1016/S2221-1691(12) 60494-8.

Barbieri A. M., Sha Q., Bette-Bobillo P., et al., 2001. "ADP-Ribosylation of Rab5 by ExoS of Pseudomonas aeruginosa affects endocytosis. *Infect Immun.*" doi: 10.1128/IAI.69.9.5329-5334.2001.

Bigger, J. W. 1994."The bactericidal action of penicillin on staphylococcus pyogenes." *Ir J Med Sci.* doi: 10.1007/BF02948386.

Borkar, D. S., N. R. Acharya, C. Leong, et al., 2014. "Cytotoxic clinical isolates of *Pseudomonas aeruginosa* identified during the steroids for corneal ulcers trial show elevated resistance to fluoroquinolones." *BMC Ophthalmol.* doi: 10.1186/1471-2415-14-54.

Deretic V., Levine B. 2009."Autophagy, Immunity, and Microbial Adaptations." *Cell Host Microbe.* doi: 10.1016/j.chom.2009.05.016.

Engel LS, Hill JM, Caballero AR, et al., 1998." Protease IV, a unique extracellular protease and virulence factor from Pseudomonas aeruginosa." *J Biol Chem.* doi: 10.1074/jbc.273.27.16792.

Feltman, H., Schulert, G., Khan, S., et al., 2001."Prevalence of type III secretion genes in clinical and environmental isolates of Pseudomonas aeruginosa." *Microbiol-SGM.* doi: 10.1099/00221287-147-10-2659.

Finck-Barbançon, V., Goranson, J., Zhu, L., et al., 1997."ExoU expression by Pseudomonas aeruginosa correlates with acute cytotoxicity and epithelial injury." *Mol Microbiol.* doi: 10.1046/j.1365-2958.1997. 4891851.x.

Fleiszig SM1, Wiener-Kronish JP, Miyazaki H et al., 1997."Pseudomonas aeruginosa-mediated cytotoxicity and invasion correlate with distinct genotypes at the loci encoding exoenzyme S." *Infect Immun.* doi: 10.1128/iai.65.2.579-586.1997.

Fleiszig, S. M. J., Zaidi, T. S., Preston, M. J., et al., 1996."Relationship between cytotoxicity and corneal epithelial cell invasion by clinical isolates of Pseudomonas aeruginosa." *Infect Immun.* doi: 10.1128/iai. 64.6.2288-2294.1996.

Fleiszig, S. M., Zaidi, T. S., Pier, G. B. 1995."Pseudomonas aeruginosa invasion of and multiplication within corneal epithelial cells in vitro." *Infect. Immun.*

Flemming H. C., Wingender J. 2010. "The biofilm matrix." *Nat Rev Microbiol.*

Frithz-Lindsten, E., Du, Y., Rosqvist, R. & Forsberg, A. 1997. "Intracellular targeting of exoenzyme S of Pseudomonas aeruginosa via type III dependent translocation induces phagocytosis resistance, cytotoxicity and disruption of actin microfilaments." *Mol Microbiol.* doi: 10.1046/ j.1365-2958.1997.5411905.x.

Fukumoto R., Cary L. H., Gorbunov N V., et al., 2013. "Ciprofloxacin Modulates Cytokine/Chemokine Profile in Serum, Improves Bone Marrow Repopulation, and Limits Apoptosis and Autophagy in Ileum after Whole Body Ionizing Irradiation Combined with Skin-Wound Trauma." *PLOS ONE.* doi: 10.1371/journal.pone.0058389.

Goodman, A. L. and Lory, S. 2004. "Analysis of regulatory networks in Pseudomonas aeruginosa by genome wide transcriptional profiling." *Curr Opin Microbiol.* doi: 10.1016/j.mib.2003.12.009.

Green M., Apel A., Stapleton F. 2008. "Risk factors and causative organisms in microbial keratitis. *Cornea.*" doi: 10.1016/j.enzmictec. 2006.07.021.

Hauser AR. The type III secretion system of Pseudomonas aeruginosa: infection by injection. 2009. *Nat Rev Microbiol.* doi: 10.1038/ nrmicro2199.

Heimer SR, Evans DJ, Stern ME, et al., 2013. "Pseudomonas aeruginosa Utilizes the Type III Secreted Toxin ExoS to Avoid Acidified Compartments within Epithelial Cells." *PLoS One.* doi: 10.1371/ journal.pone.0073111.

Hobden J. A. 2002. "Pseudomonas aeruginosa proteases and corneal virulence." *DNA Cell Biol.* doi: 10.1089/10445490260099674.

Jacoby, G. A. 2013. "Mechanisms of resistance to quinolones." *Clin Infect Dis.* doi: 10.1086/428052.

Junkins RD, Shen A, Rosen K, et al., 2013. "Autophagy enhances bacterial clearance during P. aeruginosa lung infection." *PLoS One.* doi: 10.1371/journal.pone.0072263.

Kathirvel K, Thirumalmuthu K, et al., 2020; *Comparative Gemomics of coular Pseudomonas aeruginosa strains from keratitis Pateints with different clinical outcomes;* doi: org/10.1101/2020.05.15.097220.

Keay, L., Edwards, K., Naduvilath, T., et al., 2006. "Factors affecting the morbidity of contact lens-related microbial keratitis: a population study." *Invest Ophthalmol Vis Sci.* doi: 10.1385/CBB:37:1:37.

Keren I, Kaldalu N, Spoering A, et al., 2004. "Persister cells and tolerance to antimicrobials." *FEMS Microbiol* Lett. doi: 10.1016/S0378-1097(03)00856-5.

Kim JJ, Lee HM, Shin DM et al., 2012. "Host cell autophagy activated by antibiotics is required for their effective antimycobacterial drug action." *Cell Host Microbe.* doi: 10.1016/j.chom.2012.03.008.

Krachmer JH, Mannis MJ, Holland EJ, et al., 1997. *"Cornea-cornea and external disease: clinical diagnosis and management."* St. Louis: Mosby.

Lakshmi Priya, J, Prajna, L., Mohankumar, V. 2015. "Genotypic and phenotypic characterization of Pseudomonas aeruginosa isolates from post-cataract endophthalmitis patients." *Microb. Pathog.* doi: 10.1016/j.micpath.2014.11.014.

Lalitha P, Prajna NV, Manoharan G, et al., 2015. "Trends in bacterial and fungal keratitis in South India." *British Journal of ophthalmology.* doi: 10.1186/1471-2415-14-54.

Lau GW, Hassett DJ, Britigan BE.2005."Modulation of lung epithelial functions by Pseudomonas aeruginosa." *Trends in microbiology.* doi: 10.1186/s13020-019-0250-0.

Lee, J. K., Y. S. Lee, Y. K. Park, and B. S. Kim. 2005. "Alterations in the gyrA and gyrB subunits of topoisomerase II and the parC and parE subunits of topoisomerase IV in ciprofloxacin-resistant clinical isolates of Pseudomonas aeruginosa." *Int J Antimicrob Agents*. doi: 10.1002/ptr.1695.

Levine B, Mizushima N, and Herbert W. Virgin. 2011. "Autophagy in immunity and inflammation." *Nature*. doi: 10.1016/j.chom.2009.05.016.

Levin-Reisman I, Ronin I, Gefen O, et al., 2017." Antibiotic tolerance facilitates the evolution of resistance." *Science*. doi: 10.1385/CBB: 37:1:37.

Lewis K. 2008. "Multidrug tolerance of biofilms and persister cells." *Curr Top Microbiol Immunol*. doi: 10.1007/978-3-540-75418-3_6.

Lewis K. 2014. "Persister cells, dormancy and infectious disease." *Nat Rev Microbiol*. doi: https://doi.org/10.1038/nrmicro1557.

Li J., Ji L., Shi W., et al., 2013. "Trans-translation mediates tolerance to multiple antibiotics and stresses in Escherichia coli." *J. Antimicrob. Chemother*. doi: 10.1093/jac/dkt231.

Mah, T. F., B. Pitts, B. Pellock, G. C. et al., 2003. "A genetic basis for *Pseudomonas aeruginosa* biofilm antibiotic resistance." *Nature*. doi: https://doi.org/10.1038/nature02122.

Mansilla-Pareja M. E. and Colombo M. I. 2013. "Autophagic clearance of bacterial pathogens: molecular recognition of intracellular microorganisms." *Front. Cell. Infect. Microbiol*. doi: 10.3389/fcimb.2013.00054.

Marquart M. E., Callaghan R. J. O. 2013. "Infectious Keratitis: Secreted Bacterial Proteins That Mediate Corneal Damage." *J Ophthalmol*. doi: https://doi.org/10.1155/2013/369094.

Masuda, N., Sakagawa, E, Ohya, S et al., 2000. "Substrate specificities of MexAB-OprM, MexCD-OprJ, and MexXY-OprM efflux pumps in *Pseudomonas aeruginosa.*" *Antimicrob Agents Chemother*. doi: 0.1128/aac.44.12.3322-3327.2000.

Mohankumar V, Ramalingam S, Chidambaranathan G, Prajna L. 2018. "Autophagy induced by type III secretion system toxins enhances clearance of Pseudomonas aeruginosa from human corneal epithelial cells." *Biochem Biophys Res Commun*.

Muñoz, R., and A. G. De La Campa. 1996. "ParC subunit of DNA topoisomerase IV of *Streptococcus pneumoniae* is a primary target of fluoroquinolones and cooperates with DNA gyrase A subunit in forming resistance phenotype. *Antimicrob Agents Chemother.*

Murakami K1, Ono T, Viducic D et al., 2005. "Role for rpoS gene of Pseudomonas aeruginosa in antibiotic tolerance." *FEMS Microbiol Lett.*

Murugan N, Malathi J, Umashankar V et al., 2016. "Unraveling genomic and phenotypic nature of multidrug-resistant (MDR) *Pseudomonas aeruginosa* VRFPA04 isolated from keratitis patient." *Microbiological Research.* doi: 10.1016/j.micres.2016.10.002.

Nouri R, Ahangarzadeh Rezaee M, Hasani et al., 2016. "The role of gyrA and parC mutations in fluoroquinolones-resistant Pseudomonas aeruginosa isolates from Iran." *Braz J Microbiol.* doi: 10.1016/j.bjm.2016.07.016.

Poole. 2011. "K. *Pseudomonas aeruginosa*: resistance to the max." *Front Microbiol.*

Ramirez, M. S., and M. E. Tolmasky. 2010. "Aminoglycoside modifying enzymes." *Drug Resist Updat.*

Renna M, Schaffner C, Brown K, et al., 2011. "Azithromycin blocks autophagy and may predispose cystic fibrosis patients to mycobacterial infection." *J Clin Invest.*

Sadikot R. T., Blackwell T. S., Christman J. W., Prince A. S. 2005. "Pathogen-host interactions in Pseudomonas aeruginosa pneumonia." *Am J Respir Crit Care Med.* 171 (11):1209-23.

Sharma, S. 2011. "Antibiotic resistance in ocular bacterial pathogens." *Indian J Med Microbiol.* doi: 10.4103/0255-0857.83903.

Shen E. P., Hsieh Y. T., Chu H. S. et al., 2015 "Correlation of Pseudomonas aeruginosa Genotype with Antibiotic Susceptibility and Clinical Features of Induced Central Keratitis." *Invest. Ophthalmol. Vis. Sci.* doi: 10.1167/iovs.14-15241.

Shigemura, K, Osawa. K, Kato. A. et al., 2015 "Association of overexpression of efflux pump genes with antibiotic resistance in Pseudomonas aeruginosa strains clinically isolated from urinary tract infection patients." *J Antibiot.* doi: 10.1038/ja.2015.34.

Stover, C. K., Pham, X. Q, Erwin, A. L. et al., 2000. "Complete genome sequence of Pseudomonas aeruginosa PAO1, an opportunistic pathogen." *Nature*. doi: 10.1038/35023079.

Subedi, A. Vijay, G. Kohli, S. et al.,2018."Comparative genomics of clinical strains of *Pseudomonas aeruginosa* strains isolated from different geographic sites." *Scientific Reports*. doi: https://doi.org/10.1038/s41598-018-34020-7.

Subedi, D., A. K. Vijay, and M. Willcox. 2017. "Overview of mechanisms of antibiotic resistance in Pseudomonas aeruginosa: an ocular perspective." *Clin Exp Optom*. doi: https://doi.org/10.1111/cxo.12621.

Suzuki T, Okamoto S, Oka N et al., 2018. *"Role of pvdE Pyoverdine Synthesis in Pseudomonas aeruginosa Keratitis, Cornea,"* 37 Suppl 1.

Thibodeaux BA, Caballero AR, Marquart ME et al., 2007. "Corneal virulence of Pseudomonas aeruginosa elastase B and alkaline protease produced by Pseudomonas putida." *Curr Eye Res*. doi: 10.1080/02713680701244181.

Thirumalmuthu K, Devarajan B, Prajna L, and Mohankumar V. 2018. "Mechanisms of fluoroquinolone and aminoglycoside resistance in keratitis-associated Pseudomonas aeruginosa." *Microb Drug Resist*. doi: 10.1089/mdr.2018.0218.

Tingpej, P., Smith, L., Rose, B., et al., 2007. "Phenotypic characterization of clonal and nonclonal Pseudomonas aeruginosa strains isolated from lungs of adults with cystic fibrosis." *J Clin Microbiol*. doi: 10.1128/JCM.02364-06.

Vázquez-Laslop N., Lee H., Neyfakh A. A. 2006. "Increased persistence in Escherichia coli caused by controlled expression of toxins or other unrelated proteins." *J Bacteriol*. 188(10):3494-7.

Viducic D., Ono T., Murakami K., et al., 2006. "Functional analysis of spoT, relA and dksA genes on quinolone tolerance in Pseudomonas aeruginosa under nongrowing condition." *Microbiol Immunol*. doi: 10.1111/j.1348-0421.2006.tb03793.x.

Wolfgang M. C., Kulasekara B. R., Liang X., et al., 2003. "Conservation of genome content and virulence determinants among clinical and environmental isolates of Pseudomonas aeruginosa." *Proc Natl Acad Sci*. doi: 10.1073/pnas.0832438100.

Yahr, T. L., Vallis, A. J., Hancock, M. K., Barbieri, J. T., and Frank, D. W. 1998. "ExoY, an adenylate cyclase secreted by the Pseudomonas aeruginosa type III system." *Proc Natl Acad Sci USA*. doi: 10.1073/pnas.95.23.13899.

Yu FY, Yao D, Pan JY et al.,2010." High prevalence of plasmid-mediated 16S rRNA methylase gene rmtB among *Escherichia coli* clinical isolates from a Chinese teaching hospital." *BMC Infect Dis*. doi: 10.1186/1471-2334-10-184.

Yuan K, Huang C, Fox J, et al., 2012. "Autophagy plays an essential role in the clearance of Pseudomonas aeruginosa by alveolar macrophages." *J Cell Sci*. doi: 10.1242/jcs.094573.

Yuk J. M. and Jo E. K. 2013."Crosstalk between Autophagy and Inflammasomes." *Mol cells*. doi/10.1007/s10059-013-0298-0.

Zhang Y. 2014. "Persisters, persistent infections and the Yin–Yang model." *Emerg Microbes Infect*.

Zolfaghar I., Evans D. J., Fleiszig S. M. 2003."Twitching Motility Contributes to the Role of Pili in Corneal Infection Caused by Pseudomonas aeruginosa." *Infection and Immunity*. doi: 10.1128/IAI.71.9.5389-5393.2003.

Chapter 6

MECHANISM OF QUORUM SENSING AND INHIBITION OF QUORUM SENSING MEDIATED VIRULENCE BY PUTATIVE NATURAL COMPOUNDS AGAINST GRAM-NEGATIVE BACTERIAL PATHOGENS

A. Annapoorani[1], R. Manikandan[1], A. Veera Ravi[2] and S. Janarthanan[1,]*

[1]Department of Zoology, University of Madras, Chennai, Tamil Nadu, India
[2]Department of Biotechnology, Alagappa University, Karaikudi, Tamil Nadu, India

---

[*] Corresponding Author's E mail: janas_09@yahoo.co.in.

ABSTRACT

Quorum sensing (QS) is the communicative language of several Gram-positive and Gram-negative bacterial pathogens, which orchestrates important regulatory mechanisms over the genes involved in various functions like symbiosis, virulence, competence, conjugation, antibiotic production, motility, sporulation and biofilm formation. Of which, virulence factors production and biofilm formations are the two major pivotal regulatory networks required for the progression of disease outbreaks in human as well as in aquatic organisms. With this background, this chapter depicts the QS mechanism of Gram-negative bacterial pathogens, different strategies of quorum sensing inhibitions (QSI) and their inhibitory actions through putative natural compounds. Additionally, it describes about the host-pathogen relationship and advanced computational approach to identify more novel QSIs.

Keywords: quorum sensing, bacteria, natural compounds, quorum sensing inhibition

1. INTRODUCTION

A query recurrently raised among microbiologists and biochemists during 1960s was the function that controls the light emission of "free-living" planktonic bioluminescent bacteria (Nealson et al., 1968; Kempner and Hanson, 1968). The answer was established in 1970s that there are no functions of bioluminescence when bacteria exist in planktonic state at low cell density in seawater. Those cells are unable to produce luciferase and fail to emit light (Kempner and Hanson, 1968). Further, it was explained that like other creatures, bacteria can communicate with each other and at high cell density it regulates the expression of genes with various functions (Nealson, 1977; Eberhard et al., 1981). The first bacterial communication system was described in the production of bioluminescence by marine symbiotic bacterium *Vibrio fischeri* (Nealson et al., 1970). This opens a new era for bacterial cell-cell communication system and its dependent gene expression with various physiological functions. These coordinated regulations of gene expression among the individuals of inter and intra bacterial species between prokaryotes and eukaryotes occur through a well-developed mechanism termed as "Quorum Sensing" (QS) (Givskov et al., 1996; Joint et al., 2002; Ryan and Dow, 2008; Defoirdt, 2017). Both Gram-positive and Gram-negative bacteria have been found to coordinate this QS mechanism by the production of small, diffusible signal molecules called "autoinducers" (AIs). Many Gram-positive bacteria, including *Staphylococcus aureus, Streptococcus pneumoniae, S. mutans, S. gordonii, Bacillus subtilis, Listeria monocytogenes* and *Clostridium perfringens* utilize small oligopeptides or modified peptides as AIs (Autret et al., 2003; Waters and Bassler, 2005; Ohtani et al., 2009). In Gram-negative bacteria, the most well studied QS circuit is LuxI-LuxR homologous system and cognate signal molecules are N-acyl-homoserine lactones (AHLs) (Smith and Iglewski, 2003). AHLs are known to play a key role in the production of virulence factors (Bosgelmez, 2003) and secondary metabolites (Latifi et al., 1995). This AHL mediated QS system consists of four signal components such as AHL molecule, LuxI synthase, LuxR receptor proteins as well as target genes. The AHL synthase gene is responsible for the production of AHL signal molecules. These AHLs are diffused out of the bacteria and reaches the surrounding environment. At high cell density, AHL also reaches certain threshold concentration and interacts with a cognate receptor protein (LuxR-type receptor), which in turn activate a positive transcription factor and modulates the expression of QS regulated genes (Waters and Bassler, 2005), involving symbiosis, virulence, competence, conjugation, antibiotic production, motility, sporulation and biofilm formation (Miller and Bassler, 2001) (Figure 1). This type of AHL mediated QS systems are found in many Gram-negative bacterial species belonging to the α, β, and γ subclasses of proteobacteria including genera of *Agrobacterium, Aeromonas, Burkholderia, Chromobacterium, Citrobacter, Escherichia, Enterobacter, Erwinia, Hafnia, Nitrosomonas, Obesumbacterium, Pantoea, Proteus, Pseudomonas, Rahnella, Klebsiella, Ralstonia, Rhodobacter, Rhizobium, Serratia, Vibrio, Xenorhabdus* and *Yersinia* (Eberl, 1999; Bosgelmez, 2003).

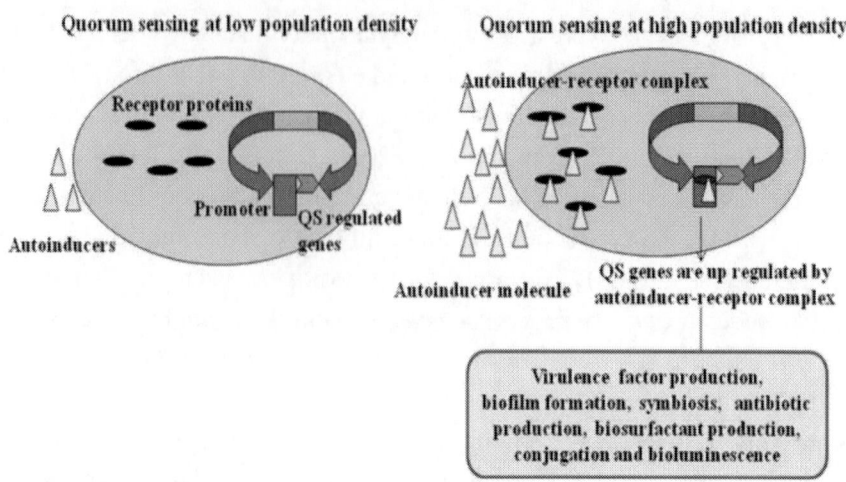

Figure 1. Mechanism of QS in Gram-negative bacteria.

1.1. Well-Known Gram-Negative Bacteria with Quorum Sensing Activity

The QS system has been well described in only few Gram negative bacteria and many of the bacterial species QS still remains elusive. Hence, this chapter emphasize about the well defined QS network of multidrug-resistant lung pathogen, urinary pathogen and aquatic pathogen such as *P. aeruginosa, S. marcescens* and *V. harveyi*, respectively.

1.2. *Pseudomonas aeruginosa*

In *P. aeruginosa*, the QS mechanism influences the expression of number of genes responsible for virulence factors production such as LasA protease, LasB elastase, pyoverdin and biofilm formation (Antunes et al., 2010). These virulence factors are involved in the development of diseases like nosocomial pneumonia, hospital-acquired urinary tract infections, bloodstream infections; severe burn wound infections in patients that cause death due to chronic lung and respiratory disorders (Van Delden and

Iglewski, 1998). The expression of virulence genes in this pathogen is mainly under the control of two well known QS systems namely, LasIR and RhlIR mechanisms (Smith and Iglewski, 2003). The Las system comprises of *lasI* synthase gene which is responsible for synthesis of N-(3-oxododecanoyl)-L-homoserine lactone (3-oxo-C12-HSL) and transcriptional regulator LasR (Whitehead et al., 2001). This receptor and signal complex regulates the production of elastase, exotoxin A and alkaline protease (Senturk et al., 2012). The Rhl system is relying on N-butanoyl-L-homoserine lactone (C4-HSL). This C4-HSL production is under the control of *rhlI* synthase gene and form a complex with RhlR that leads to the expression of rhamnolipid, alkaline protease, elastase, cyanide and pyocyanin (Bosgelmez, 2003). However, the LasIR system, in turn regulates the expression of RhlIR system (Schuster and Greenberg, 2006). In addition to these two QS systems, a third system of QS is also reported as *Pseudomonas* Quinolone Signal (PQS) (Pesci et al., 1999). PQS signal belongs to the family of 4-hydroxy-2-alkylquinolines (HAQs) under the control of transcriptional regulator called MvfR. There are two operons such as *pqsABCDE* and *phnAB* were found to implicate in the biosynthesis of HAQs. This positively controls the expression of many genes involved in the synthesis of anthranilic acid and its conversion to 4-hydroxy-2-heptylquinoline (HHQ) (Deziel et al., 2004; Sifri, 2008). Further, HHQ is converted to 2-heptyl-3-hydroxy-4-quinolones by the action of PqsH, which is also known as PQS (Deziel et al., 2004). It is well known that the PQS, Las and Rhl systems are interdependent to each other (Smith and Iglewski, 2003). Consequently, it is well documented that the *las*, *rhl* and *mvfR* genes of QS systems are components of a large network of regulators that control a wide assortment of cellular functions and pathogenesis of *P. aeruginosa* (Sifri, 2008). In addition to this, other global regulators such as *gacA*, *rpoS* and *rpoN* have also been demonstrated to regulate the expression of virulence factors (Suh et al., 1999; Parkins et al., 2000). To facilitate the establishment of infection, *P. aeruginosa* produces aforementioned QS dependent virulence factors and biofilm formation (Van Delden and Iglewski, 1998; Bosgelmez, 2003) (Figure 2).

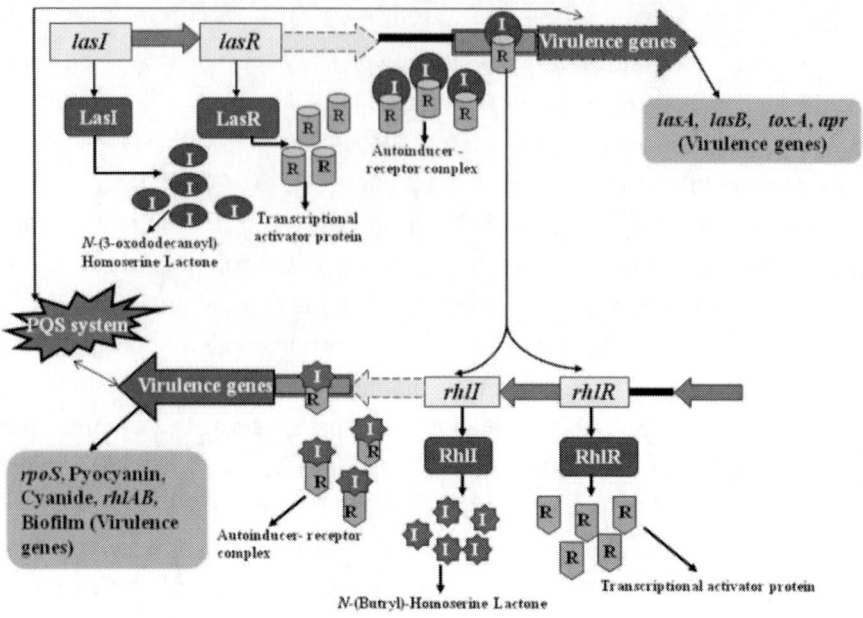

Figure 2. QS mechanism in *P. aeruginosa*.

1.3. *Serratia marcescens*

The *Serratia* spp. are motile, facultative anaerobic, chemoorganotrophic bacteria with both respiratory and fermentative type of metabolism (Houdt et al., 2007). Among many Gram-negative pathogenic bacteria, *S. marcescens* is one of the most common pathogens in humans, which leads to nosocomial infections, urinary tract infections, respiratory system infections, dermatitis, conjunctivitis, keratitis, endophthalmitis, intravenous catheter-associated infections and chronic wound infections (Hejazi and Falkiner, 1997; Stock et al., 2003). Hence, controlling the virulence of *S. marcescens* infections in clinical patients is one of the biggest issues in medicine and microbiology (Arakawa et al., 2000). It has been known that *S. marcescens* QS mechanism governs the expression of secondary metabolites such as prodigiosin, carbapenem, virulence factors such as nuclease, protease, lipase, hemolysin and biofilm-related behaviours of *S. marcescens* (Sarah et al., 2006; Wei and Lai, 2006). Different types of AHLs production such as N-hexanoyl-L-homoserine lactone (C6-HSL) and C4-HSL are the most common AHLs identified in *Serratia* spp. (ATCC 39006) and *S. marcescens* strains MG1. C6-HSL, N-heptanoyl-L-homoserine lactone (C7-HSL), N-octanoyl-L-homoserine lactone (C8-HSL) and N-3-oxo-hexanoyl-L-homoserine lactone (3-oxo-C6-HSL) are known to be produced by *S. marcescens* SS-1 and 3-oxo-C6-HSL mediated QS gene expression has also been reported in *S. proteamaculans* B5a (Wei and Lai, 2006; Houdt et al., 2007) (Figure 3).

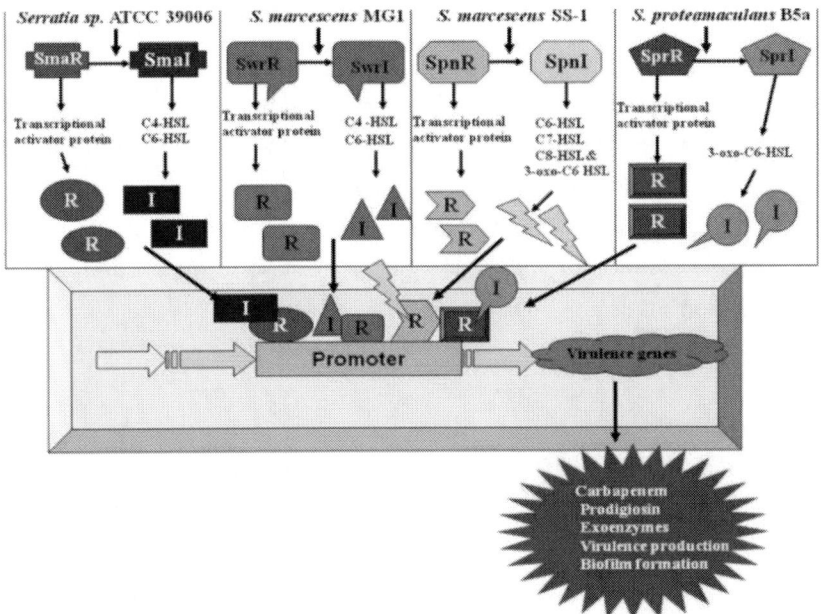

Figure 3. QS mechanism in *S. marcescens*.

1.4. *Vibrio harveyi*

Vibrio harveyi is a most common, marine Gram-negative luminous bacteria capable of infecting wide range of marine organisms including fishes and many invertebrates particularly penaeid shrimps (Austin and Zhang, 2006). There is well described three channel QS system of *V. harveyi* viz., AHL mediated harveyi autoinducer 1 (HAI-1), furanosyl borate diester mediated autoinducer 2 (AI-2) and cholerae autoinducer 1 (CAI-1) (Defoirdt et al., 2008). LuxM, LuxS, and CqsA enzymes are known to be responsible for the production of HAI-1, AI-2, and CAI-1, respectively. The two-component receptor proteins such as LuxN, LuxQ and CqsS are required for the cognate binding of these AIs, respectively. The phosphorylation and dephosphorylation signal transduction cascade is mediated through LuxO protein. LuxO requires periplasmic protein LuxP for the recognition of AI-2 molecules. In the absence of AI, the receptors will autophosphorylate which transfers the phosphate to LuxO via LuxU. Phosphorylation activates the LuxO, which together with σ^{54} activates the five small regulatory syntheses of RNAs (sRNAs). These sRNAs along with the chaperone Hfq, destabilize the mRNA of transcriptional regulator LuxR. Hence, LuxR protein is not produced in the absence of AI. If the AI concentration increases, then the receptor protein exchange from kinases to phosphatases. The dephosphorlation of LuxO inactivates the synthesis of sRNAs which leads to the activation of LuXR. Further, the active form of LuXR triggers the expression of target genes (Fig. 4).

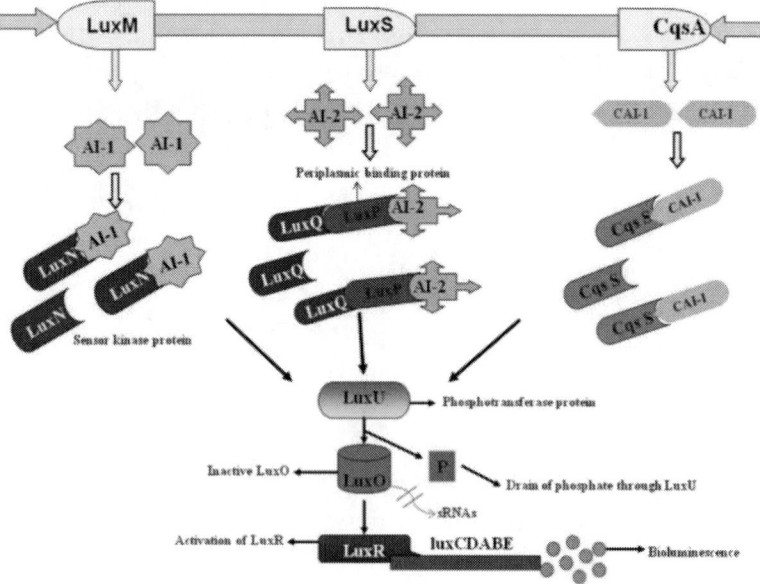

Figure 4. QS mechanism in *V. harveyi* at high cell density.

2. Antibiotic Resistance and Bacterial Biofilm

The discovery of penicillin by Alexander Fleming in 1928 has shown a broad range of efficacy against bacterial pathogens through growth inhibitory mechanism. Consequently, the development of antimicrobial pharmaceuticals was found to be the only way for the treatment of infectious diseases (Hentzer and Givskov, 2003). However, the indiscriminate use of antibiotics resulted in frequent bacterial mutations which are often caused an increased incidence of antibiotic resistance (Hentzer and Givskov, 2003). The antibiotic resistance spreads by the way of horizontal transfer of resistance genes among the bacterial population and it is achieved either by degradation of antibacterial compounds or decreased permeability or affinity for antibiotics or leads to efflux of these molecules (Hancock, 1998; Lewis, 2001). Thus, the traditional way of antimicrobial treatment becomes ineffective due to the increased occurrence of multidrug-resistance among the pathogenic bacteria (Lewis, 2001). Biofilm is a complex aggregation of microbial cells in the self secreted exopolymeric matrix. The microbial cells found inside the biofilm are well protected by negatively charged exopolysaccharides (EPS) and thereby restrict the permeation of positively charged antibiotics possibly through binding (Shigeta et al., 1997; Ishida et al., 1998). The incapable nature of antimicrobials penetrate through biofilm possibly resulted in the expression of resistance genes (Lewis, 2001). It is also known that antibiotics like fluoroquinolones could effectively equilibrate across the biofilm and inhibit the growth of bacteria (Anderl et al., 2000). Even in such situation, the biofilm formation retards the diffusion of antimicrobial substances and decreases the concentration of antibiotics entry and thereby reduces the efficiency of concerned antibiotic molecule. Further, the production of hydrolytic enzyme namely β-lactamase destroys the incoming antibiotics (Giwercman et al., 1991; Stewart, 1996). Another interesting diffusion barrier is production of catalase in *P. aeruginosa* in protecting biofilm against hydrogen peroxide (H_2O_2) generation. In such situation, the inducible activities of katB producing catalase protect the cells inside the biofilm effectively from the entry of H_2O_2. The restricted penetration and destruction of antimicrobial substances by microbial cells confer the expression of more resistance genes (Nikaido, 1994) (Figure 5). In order to overcome the complications involved in the use of antibiotics with reduced efficacy to fight against potent bacterial pathogens, modern medicine is entering into a post-antibiotic era. This makes the pressure to develop novel alternative therapeutic approaches (Hentzer and Givskov, 2003).

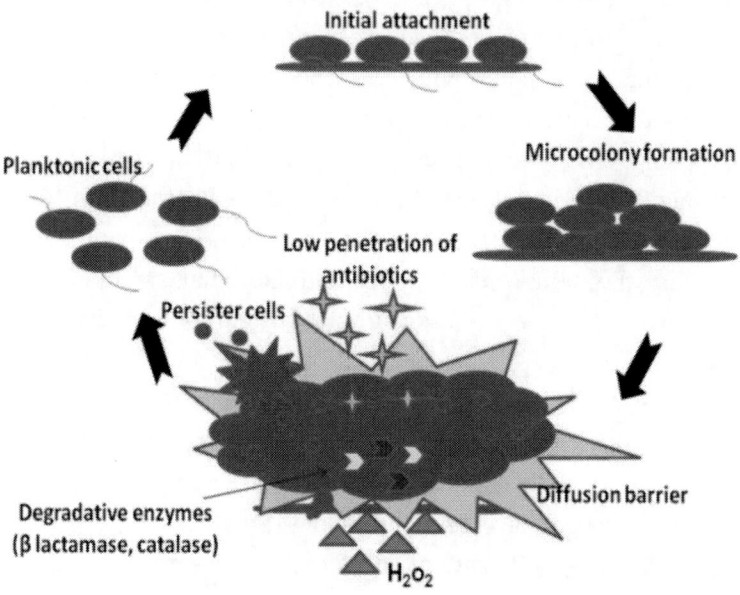

Figure 5. Developmental stages of biofilm and its resistance mechanism.

2.1. Bacterial Interaction in Dental Biofilm Development

The diversity of oral bacterial species plays a vital role in the maintenance of oral health and disease in humans (Kumar et al., 2005). The dental bacterial communities form a sequential colonization of dynamic structural complex on the tooth enamel is called dental biofilms (Kolenbrander et al., 2002). Human population worldwide are heavily affected by the oral diseases related to dental biofilms (Koo, 2008). The constituents of diet like sucrose and starch are fermentable as well as serve as a substrate for the synthesis of EPS and intracellular polysaccharides (IPS) in dental biofilm, thus resulted in the development of dental plaque on enamel surfaces (Bowen, 2002). In addition, the secondary metabolites of several oral bacteria present in the dental biofilm are used as an energy source for the bordering bacteria (Hojo et al., 2009). Subsequently, the primary step for the proteinaceous film development on the tooth surfaces is due to the adsorption and adhesion of certain microbial species present in the dental cavity, which are known as early colonizers (Yao et al., 2003). The early colonizers pave the way for the attachment of subsequent colonizers, which influence the succeeding stages of biofilm formation (Li et al., 2004). These dental colonizers rely on various physical and nutritional factors, in which, QS is one among the factors to accelerate their cellular behavior on the dental hard tissues (Hojo et al., 2009). AI-2 and competence-stimulating peptide (CSP) mediated signaling are observed in most of the early colonizers of dental biofilm such as *Porphyromonas gingivalis, S. mutans, S. gordonii* and *S. intermedius* (Hojo et al., 2009). The bacterial cells reside inside the dental biofilm are frequently exposed to various stress conditions like extreme nutrient storage, low pH, high osmolarity, oxidation, and consumption of high dosage of antibiotics. The bacterial colonizers in such conditions are observed to adapt to this new environment and become resistance to those stress conditions (Donlan and Costerton, 2002).

3. Host - Gram-Negative QS Pathogen Interaction

The pathogenesis and the development of disease condition depend on the efficiency of bacterial community and host innate immune system (Diacovich and Gorvel, 2010). Using QS, bacteria can synchronize their community behavior in terms of manipulating the host cell activities even in a noninvasive manner. The early stage of infection occurred between the bacteria and host cells may takes place even before the binding of bacteria enter into the cell and then promote distribution inside the host (Vikstrom,

2019). The high concentration of AHL plays an important role in modulating the host innate immune response through several cellular processes such as homeostasis, mitochondrial and cytoskeletal dynamics, calcium signaling, governing transcriptional and proteomic responses of host cells (Peterson and Artis, 2014; Vikstrom et al., 2010).

Diffusion of QS signal molecules across the plasma membrane or the membrane of the sub cellular organelles differs based on their efficacy. For example, C4 BHL of *P. aeruginosa* can easily get diffused across the membrane. In contrast, the long chain of 3O-C12-HSL requires interaction of phospholipids, membrane micro domains such as caveolae and lipid rafts for the entry of long chain into the membrane (Vikstrom, 2019). To facilitate this process, interaction with sub membrane organelles such as filamentous actin, actin-binding proteins and anchoring membrane proteins are essential for the successful invasion of long acyl chain AHLs. Along with this, interaction with IQ-motif-containing GTPase-activation protein (IQGAP1) is a human target for 3O-C12-HSL. IQGAP1 and phosphorylation of Rac1 and Cdc42 are the most important changes in the actin cytoskeleton network (Karlsson et al., 2012). Using biotin-conjugated 3O-C12-HSL probe and mass spectrometry-based proteomic analysis, IQGAP1 was proven as a human target for bacterial 3O-C12-HSL (Karlsson et al., 2012).

The *P. aeruginosa* 3O-C12-HSL activates by the linkage of chemosensory G-protein-coupled receptor T2R38 on leukocytes and also binds to nuclear peroxisome proliferator-activated receptors (PPAR). This interactions lead to the modulation of DNA binding activity and transcription of NF-κB dependent genes (Jahoor et al., 2008). At last, the long chain acyl 3O-C12-HSL signal molecule passing through the plasma membrane and reaches the host cells. These interaction of long chain AHL with many intracellular molecules and signaling cascades, leads to the modulation of RNA and DNA processes in eukaryotic cells (Vikstrom, 2019).

4. ROLE OF QS MEDIATED VIRULENCE DURING PATHOGENESIS

The virulence factors produced by bacterial pathogens are known to alter the equilibrium of host defence mechanism. After colonization into the host, the bacteria secrete several extracellular virulence factors such as enzymes and toxins, which cause extensive tissue damage, dissemination, systemic inflammatory-response syndrome and death (Van Delden and Iglewski, 1998). Production of many of the virulence factors are under the control of QS. In case of virulence in *P. aeruginosa*, the exotoxin A is involved in the inhibition of host cell protein synthesis by means of inactivating the elongation factor-2 and cause cell death (Wick et al., 1990), Further, direct tissue damage of chronic lung infection, abnormal mucociliary function of respiratory epithelium (Rahim et al., 2001) and corneal infections are caused by the exoenzyme S, rhamnolipid and alkaline proteases of *P. aeruginosa* (Kernacki et al., 1995), respectively. LasB elastase of *P. aeruginosa* was observed to destruct the major part of human lung tissue and joints made up of elastin responsible for muscle contraction, expansion and also damage blood vessels that lead to pulmonary haemorrhages (Galloway, 1991). In addition, LasA protease, a well known serine protease that acts synergistically with LasB elastase and causes extensive tissue damage (Kessler et al., 1982). Similarly, *S. marcescens* also plays a vital role in the infectious diseases through mechanisms mentioned earlier. This pathogen secretes virulence factors like nuclease, protease, lipase, chitinase, cellulase and hemolysin for the progression of various human infectious diseases (Hejazi and Falkiner, 1997). Likewise, *Vibrio* spp. is responsible for the mortality of various aquaculture organisms. Several QS mediated virulence factors like extracellular toxin, metalloprotease, type III secretion system and siderophore production cause disease outbreaks in aquaculture organisms (Defoirdt et al., 2008).

Since the QS plays a vital role in the production of virulence factors and biofilm formation, the inhibition of such QS mechanism could be adopted as an alternative strategy to prevent the development of bacterial resistance and makes the pathogen become ineffective to establish successful infection without imposing any selective pressure on bacterial growth (Hentzer and Givskov, 2003). This opens a new avenue for the entry of quorum sensing inhibitor (QSI) compounds to treat bacterial infections.

5. TYPES OF QSI

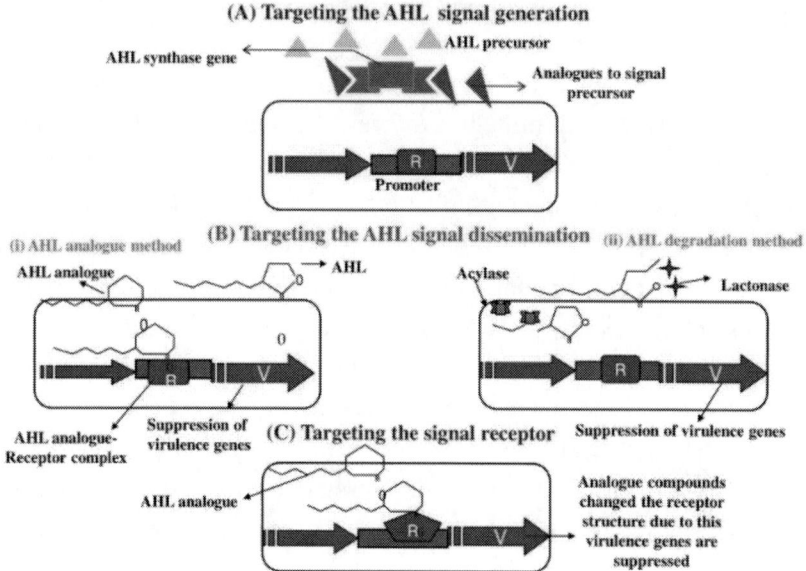

Figure 6. Types of QSI mechanism. (A) Targeting the AHL signal generation represents the binding of AHL analogue compounds instead of AHL precursors with the AHL synthase gene prevents the synthesis of AHL signal molecule. (B) Targeting the AHL signal dissemination represents two types of inhibitory mechanisms. (i) Antagonistic binding of AHL analogues instead of AHL signals and suppress the virulence gene expression. (ii) Degradation of AHL signals by acylase and lactonase enzymes, thereby inhibits the virulence genes expression. (C) Targeting the signal receptor represents the binding of AHL analogues alters the structure binding domain or receptor protein. Thereby it inhibits the expression of virulence genes.

The QSIs could target the QS system of bacterial pathogens by three ways viz., (1). Inhibition of AHL synthesis, (2). Degradation of secreted AHL molecules and (3). The competitive binding of AHL analogue with receptor proteins (Rasmussen and Givskov, 2006) (Figure 6).

5.1. Inhibition of AHL Synthesis

It is well known that the interference of AHL production modulates the QS pathway (Rasmussen and Givskov, 2006). Targeting of LuxI type synthase proteins using small molecules affect the synthesis of AHL (Geske et al., 2008). The LuxI type synthases mediated AHL production generally occur through sequential reaction mechanism by utilizing S-adenosylmethionine (SAM) as amino donor for the formation of homoserine lactone ring moiety and acyl carrier protein (ACP) as the precursor to acyl side chain (Schaefer et al., 1996). The binding of SAM and acyl ACP with AHL synthase gene activate the acylation and lactonization reactions (Parsek et al., 1999; Hentzer and Givskov, 2003; Rasmussen and Givskov, 2006). Finally, the AHL is released along with the by-product of holo-ACP and 5'-methylthioadenosineis. For example, triclosan was found to inhibit the ACP reductase and thereby inhibited the synthesis of AHL. Various analogues of SAM like S-adenosyl-homocysteine, butyryl SAM and sinefungin are proven to be an effective inhibitors of *rhlI* synthase gene of *P. aeruginosa* (Hentzer and Givskov, 2003; Rasmussen and Givskov, 2006). These analogue molecules bind to the synthase gene instead of SAM and thereby inhibited the synthesis of AHL. Other AI blocking compounds like thiol derivatives and alkylated thio derivatives could also be used as analogues of purine nucleotide and homoserine lactone derivatives (Cronan et al., 2000) (Figure 6).

5.2. Degradation of AHL Molecules

The inactivation or complete degradation of secreted signal molecules could be achieved by various methods such as chemical degradation, enzymatic destruction interfering AHL synthesis pathway and degrading AHL molecules at alkaline pH thereby prevent the expression of virulent genes. Similarly, high temperature is one among the factors that play a role in AHL ring-opening, but it is counteracted by the length of side chains, which decreases the rate of lactonolysis (Rasmussen and Givskov, 2006). In addition, enzymes like lactonase and acylase play a vital role in AHL degradation by the cleavage of lactone ring and acyl side chains, respectively (Figure 6). This lactonase enzyme is produced by *B. cereus*, *B. mycoides* and *B. thuringiensis*. A gene called *aiiA* in these bacterial organisms is responsible for the synthesis of lactonase enzyme (Dong et al., 2000). Besides, several other bacterial spp. such as *P. aeruginosa* PAI-A, *Rhodococcus* spp. (Uroz et al., 2003), *K. pneumoniae*, *A. tumefaciens* (Carlier et al., 2003) and *Anthrobacter* spp. (Park et al., 2003) were also been reported to produce AiiA homologs. Similarly, inactivation of AHLs by *aiiD* gene through AHL acylase synthesis by breaking the amide bond present in acyl side chains of the AHLs and produce fatty acid plus homoserine lactone ring as the by-products is also reported (Lin et al., 2003). *Variovorax paradoxus* (Leadbetter and Greenberg, 2000), *P. aeruginosa* (Sio et al., 2006), *Ralstonia* spp. (Lin et al., 2003) and *Rhodococcus erythropolis* (Uroz et al., 2005) are known to produce AHL acylase.

5.3. Competitive Binding of AHL Analogues

Blocking of receptor proteins with AHL analogue compounds is widely used method of QSI. Certain inhibitor compounds modulate the QS pathway through the interaction with LuxR type receptor protein. These types of compounds have been discovered using the known structure of AHL signaling molecule as a template (Galloway et al., 2011). These compounds make either competitive binding with the receptor or prevent the signal recognition or extreme changes in the protein folding could degrade the receptor protein. This binding normally resulted in the prevention of the expression of virulence and biofilm responsive genes (Figure 6). Several antagonistic type of QSI compounds have been reported by a group of researchers. For example, furanone is known to affect the QS system in terms of reducing the half-life of LuxR receptor protein (Manefield et al., 2002). Similarly, patulin from penicillium species inhibit QS by affecting the LuxR receptor protein (Rasmussen et al., 2005). In case of Gram-positive bacteria, autoinducing peptide (AIP) antagonists inhibit QS mechanism by preventing the binding of AIP to their respective receptor protein (Ni et al.,

2009). Similarly, most of the QSI compounds such as 4-nitro-pyridine-N-oxide (4-NPO) (Rasmussen et al., 2005), phenyl acetic acid (Musthafa et al., 2012), and methyl eugenol (Packiavathy et al., 2012a), marine sponges such as *Aphrocallistes bocagei, Haliclona (Gellius) megastoma* and *Clathria atrasanguinea* (Annapoorani et al., 2012a), mangroves such as *Rhizophora apiculata* and *R. mucronata* (Annapoorani et al., 2013), have proven to be potential antagonistic.

6. NATURAL RESOURCES OF QSIS

Natural products are believed to be a prospective source of nutraceuticals or pharmaceuticals to discover new lead compounds for the treatment of human diseases. The ethnopharmacological and traditional medicine studies have provided numerous drugs to the international pharmacopeia. The traditional natural medicines act as a major source for the development of new lead compounds for many pharmaceutical industries (Patwardhan et al., 2004). Moreover, the problems dealing with drug-resistant microorganisms, side effects of modern drugs and emerging diseases have stimulated interest in exploring dietary plants as a significant source of new medicines. Ayurvedic medicinal plants have widely been used for pharmacognosy, pharmacology and clinical therapeutics research (Dahanukar et al., 2000). Numerous alkaloids like rauwolfia and holarrhena were used to treat hypertension (McGregor and Segel, 1955) and amoebiasis (Acton and Chopra, 1933), respectively. Phenolic and flavonoid compounds like carotenoids and fucoxanthin were used to treat various diseases like cancer (Spirt et al., 2010), cardiovascular ailments (Riccioni et al., 2008), muscular degeneration (Snodderly, 1995) and also possesses numerous biological activities like anti-inflammatory (Shiratori et al., 2005), antidiabetic (Maeda et al., 2007) and antioxidant properties (Sachindra et al., 2007). Pharmaceutical industries already launched the market for vincristine and vinblastine several million dollars worth to treat childhood leukemia and Hodgkin's disease. The US National Cancer Institute has assigned to screen 50,000 natural substances for the anti-cancer property against the cancer cell lines and also for anti-viral activity against Human Immunodeficiency Virus (HIV) (Patwardhan et al., 2004). Natural product research has been used to discover new lead structures, which can be used as a template for the discovery of new drugs by the pharmaceutical industries. Though, the natural compounds are known for their great deal of bioactive potential, studies on their QSI activity are very much limited till date. Moreover, it is envisaged that the plants surviving in a natural environment should possess natural protective agent to fight against bacterial infections (Cos et al., 2006). Hence, researchers increasingly search for new therapeutic and non-toxic inhibitors of QS from natural resources (Hentzer and Givskov, 2003; Choo et al., 2006; Adonizio et al., 2008; Khan et al., 2009; Musthafa et al., 2010; Annapoorani et al., 2012a, 2012b and 2013; Packiavathy et al., 2012a). It is stipulated that an effective and applicable QSI compound should possess low molecular mass, high degree of specificity for the QS regulator, stability, resistant to metabolism and easy disposal by the host without any toxic side effects (Rasmussen and Givskov, 2006). Though, the plant-derived phytochemicals are considered to be the richest reservoir for the new and novel therapeutics (Kumar et al., 2006), only a limited number of compounds were identified with the QSI potential. Hence, a profound investigation is required to explore the non-toxic and stable QSI compounds from the plant resources to act against the multiple AHL mediated QS systems of Gram- negative bacterial pathogens.

Furanone was identified from red marine alga *Delisea pulchra* as the very first QSI compound (Manefield et al., 1999). The marine alga *Laminaria digitata* is known to possess oxidized halogen compounds help in preventing the biofouling on the surface of this marine alga. These compounds are capable of inactivating the AHL and thereby interfere with the QS controlled gene expression of colonizing bacteria (Borchardt et al., 2001). The sub-inhibitory concentration of cinnamaldehyde isolated from a common spice interfered with the QS systems of *V. harveyi* and on the biomarker strain *E. coli* (Niu et al., 2006). The extract of *Vanilla planifolia* showed QSI activity against *C. violaceum* CV026 (Choo et al., 2006). The compounds betonicine, floridoside and isethionic acid from red alga *A. flabelliformis* inhibited the C8-HSL mediated gene expression in *A.*

tumefaciens (Kim et al., 2007). The extract of south Florida medicinal plants such as *Conocarpus erectus*, *Chamaecyce hypericifolia*, *Callistemon viminalis*, *Bucida burceras*, *Tetrazygia bicolor* and *Quercus virginiana* (Adonizio et al., 2008) and manoalide derivatives of marine sponge *Luffariella variabilis* (Skindersoe et al., 2008) showed potential QSI activity against various QS mediated phenomena in *P. aeruginosa*. Plant essential oils are known to have QSI potential to inhibit QS mediated phenomena in *C. violaceum* (ATCC 12472 and CV026) and *P. aeruginosa* PAO1 (Khan et al., 2009). Flavonoids like naringenin, kaempferol, quercetin and apigenin were effective antagonists against the biofilm formation in *V. harveyi* (Vikram et al., 2010). The aqueous extract of edible fruits and plants namely *Ananas comosus*, *Musa paradiciaca*, *Manilkara zapota*, and *Ocimum sanctum* exhibited QSI activity against *C. violaceum* and *P. aeruginosa* PAO1. In this case, unknown antagonistic nature of compounds significantly inhibited various QS mediated phenomena in the aforementioned bacterial strains (Musthafa et al., 2010). The common south Indian spice *Capparis spinosa* and curcumin were proven for its QSI and antibiofilm potential against various Gram-negative uropathogens such as *P. aeruginosa*, *E. coli*, *P. mirabilis* and *S. marcescens* (Abraham et al., 2011; Packiavathy et al., 2012b). Similarly, an another common south Indian spice *Cuminum cyminum* and its secondary metabolite methyl eugenol exhibited potential QSI and antibiofilm activity against various Gram-negative bacterial pathogens such as *P. aeruginosa*, *P. mirabilis* and *S. marcescens* (Packiavathy et al., 2012a), Different solvent extracts of spice clove bud namely *Syzygium aromaticum* showed QSI activity against different AHL dependent characteristics of biomarker strains *C. violaceum* (CV026), *E. coli* and *P. aeruginosa* (Krishnan et al., 2012). The derivative of organosulfur compounds such as S-phenyl-L-cysteine sulfoxide and its breakdown product, diphenyl disulfide from onion and garlic were significantly reduced the biofilm formation of *P. aeruginosa* and recovered the *Drosophila melanogaster* from the *P. aeruginosa* infection (Cady et al., 2012). The sub-inhibitory concentrations of honey affect the MvfR, Las and Rhl regulons and their associated virulence factors production of *P. aeruginosa* (Wang et al., 2012). Tasco, a product made from the brown seaweed namely *Ascophyllum nodosum* suppressed various QS genes expression such as *lasI, lasR, rhlI, rhlR, hcnC, aroE, rpoN, sbe, sodB, phz* and *pyoS3A* mediated virulence factors production in *P. aeruginosa*. This product enhanced the survival of *Caenorhabditis elegans* against *P. aeruginosa* infection (Kandasamy et al., 2012). Green cardamom against *C. violaceum* (Abdullah et al., 2017), naringin from pummelo peel flavonoid extract against *C. violaceum* and *V. anguillarum* (Liu et al., 2017), Coumarin group of compounds against several clinically relevant pathogens and also against plant pathogens (Reen et al., 2018), plant-derived molecules like pyrogallol and coumarin against *C. violaceum* (Inchagova et al., 2019), essential oils against several Gram-negative multidrug-resistant clinical isolates (Alibi et al., 2020) revealed the QSI potential of natural edible compounds. The derivatives of 3-amino-2-oxazolidinone were proven as a potent QSI against *C. violaceum* and *P. aeruginosa* PAO1. The synthesized compound also enhanced the life span of *C. elegans* N2 against PAO1 infection (Jiang et al., 2020).

7. COMPUTATIONAL APPROACH IN DRUG DISCOVERY

In the last few decades, pharmaceutical industry acquired more investment and consumed more time to bring new products to the market. The target identification is the preliminary step in the drug discovery process (Terstappen and Reggiani, 2001). Further, the identified drug target must satisfy various criteria for the progression of these drugs to enter into commercialization (Terstappen and Reggiani, 2001; Baldi, 2010). Though the conventional screening methodology is effective, the problems associated with these methods are expensive and time-consuming process (Baldi, 2010). This creates pressure to the research and development for the findings of inexpensive technology and also to shorten the length of time spent in the drug discovery process. It ends with the beginning of new era in the drug discovery process called "computer-aided drug discovery (CADD)" (Ooms, 2000). Drug discovery using CADD approach has many major advantages with that of conventional screening methodology. CADD is a widely used term that represents enormous computational tools for designing compounds with interesting physiochemical relationships as well as systemic assessment of potential lead candidates before they are synthesized and tested (Song et al., 2009). CADD now plays a critical role in the generation of vast libraries of pharmacologically important lead compounds and the development of new algorithms to evaluate the potency and selectivity of lead candidates with good adsorption distribution metabolism and excretion (ADME) properties (Klebe, 2006). The properties of an ideal drug has been well described by Mohan et al. (2012), who have mentioned that an ideal drug must be safe and effective, should be well absorbed orally and bioavailable, the drug should be metabolically stable and with a long half-life, should be non-toxic with minimal (or) no side effects and it should have selective distribution to target tissues. The foundation for the CADD was established in the early 1970s, in order to modify the biological activity of insulin using structural biology (Blundell et al., 1972). CADD process is more logical and target specific, process execution steps are parallel, compatible, drug development requires only one-third of amount spent for conventional screening, drug development requires very minimal time, easy management and the redundancy can be avoided (Baldi, 2010). The availability of pathogen genome sequences is considered to be a rich source of information to design targeted drugs (Miesel et al., 2003). Hence, the discovery of novel drugs using this CADD approach is more important to cure emerging diseases in future. The CADD based technology includes two major approaches such as structure-based drug design and ligand-based drug design.

7.1. Structure-Based Drug Design

The core objective of the structure-based drug design is to find out the lead compound that fits the best into the binding cavity of the known protein structure (Kolb et al., 2009; Baldi, 2010). Molecular docking program is used to screen structure-based small organic molecules to fit into the target protein structure. Based on the docking score, structural and chemical complementarities, the molecules are tested experimentally (Kolb et al.,

2009). After the preclinical and clinical evaluation of these drugs, the compounds become available for the treatment (Song et al., 2009) (Figure 7).

Using this technology, several successful drugs have already been discovered which includes dorzolamide for the treatment of cystoid macular edema (Grover et al., 2006), zanamivir for influenza infection (Itzstein et al., 1993) and amprenavir for the treatment of HIV infection (Goodgame et al., 2000). In addition to these drugs, several QSIs like baicalin, baicalein, esculin, esculetin and psoralen were identified against *A. tumefaciens* and *P. aeruginosa* (Zeng et al., 2008). Three lead compounds such as salicylic acid, nifuroxazide, and chlorzoxazone displayed QSI activity against LasIR and RhlIR controlled virulence factors of *P. aeruginosa* are also known (Yang et al., 2009). Similarly, Annapoorani et al., 2012b identified a few potential QSIs from natural edible fruits and plants using this CADD approach. Using this CADD, totally 1,920 natural compounds were docked against LasR and RhlR receptor proteins of *P. aeruginosa*. Based on the docking scores, five top-ranking compounds such as rosmarinic acid, mangiferin, morin, chlorogenic acid and naringin were chosen with proven QSI activity.

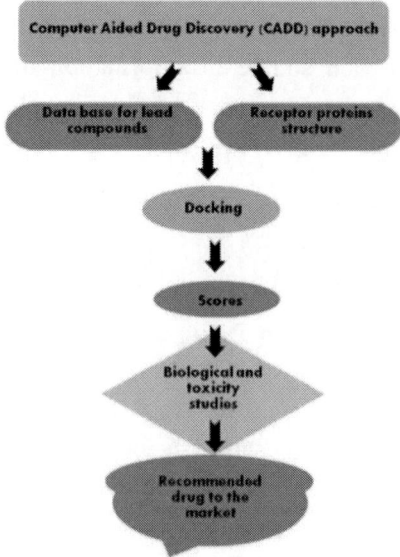

Figure 7. Schematic representation of structure based drug design.

CONCLUSION

This chapter describes the expression of QS genes in chosen Gram-negative bacterial pathogens along with the efficiency evaluation of natural compounds to inhibit the QS mediated expression of virulence and biofilm-forming genes. The discovery of more and more QSIs using conventional, as well as CADD, approaches pave the way for the entry of new QSI drugs as an alternative to antibiotics to treat multidrug-resistant bacterial pathogens.

ACKNOWLEDGMENTS

We gratefully acknowledge Department of Zoology, University of Madras (Guindy Campus), Chennai and Department of Biotechnology, Alagappa University, Karaikudi for their academic support. Dr. A. Annapoorani gratefully acknowledges Dr. D. S. Kothari Post Doctoral Fellowship, UGC, New Delhi for the financial assistance rendered [F.4-2/2006(BSR)/BL/18-19/0298].

REFERENCES

Abdullah Ali Asghar, Masood Sadiq Butt, Muhammad Shahid, and Qingrong Huang.2017. "Evaluating the Antimicrobial Potential of Green Cardamom Essential Oil Focusing on Quorum Sensing Inhibition of *Chromobacterium violaceum*." *Journal of Food Science and Technology*. 54: 2306-2315. doi.org/10.1007/s13197-017-2668-7.

Adonizio Allison, Kok-Fai Kong, and Kalai Mathee. 2008. "Inhibition of Quorum Sensing-Controlled Virulence Factor Production in *Pseudomonas aeruginosa* by South Florida Plant Extracts." *Antimicrobial Agents and Chemotherapy* 52:198-203. doi: 10.1128/AAC.00612-07.

Alibi Sana, Walid Ben Selma, Jose Ramos-Vivas, Mohamed Ali Smach, Ridha Touati, Jalel Boukadida, Jesus Navas, and Hedi Ben Mansour. 2020. "Anti-oxidant, Anti-bacterial, Anti-biofilm, and Anti-quorum Sensing Activities of Four Essential Oils against Multidrug-Resistant Bacterial Clinical Isolates." *Current Research in Translational Medicine* 68(2):59-66. doi.org/10.1016/j.retram.2020.01.001.

Ammermann Eberhard, A.L. Burlingame, Eberhard B. Cimijotti, George L. Kenyon, Kenneth H. Nealson, and N.J. Oppenheimer. 1981. Structural Identification of Autoinducer of *Photobacterium fischeri* Luciferase. *Biochemistry* 20:2444-2449. doi: 10.1021/bi00512a013.

Anderl Jeff N., Michael J. Franklin, and Philip S. Stewart. 2000. "Role of Antibiotic Penetration Limitation in *Klebsiella pneumoniae* Biofilm Resistance to Ampicillin and Ciprofloxacin." *Antimicrobial Agents and Chemotherapy* 44:1818-1824. doi: 10.1128/aac.44.7.1818-1824.2000.

Annapoorani Angusamy, Abdul Karim Kamil Abdul Jabbar, Syed Khadar Syed Musthafa, Shunmugiah Karutha Pandian, and Arumugam Veera Ravi. 2012. "Inhibition of Quorum Sensing Mediated Virulence Factors Production in Urinary Pathogen *Serratia marcescens* PS1 by Marine Sponges." *Indian Journal of Microbiology* 52:160-166. doi: 10.1007/s12088-012-0272-0.

Annapoorani Angusamy, Balaji Kalpana, Khadar Syed Musthafa, Shunmugiah Karutha Pandian, and Arumugam Veera Ravi. 2013. "Antipathogenic Potential of *Rhizophora* spp. Against the Quorum Sensing Mediated Virulence Factors Production in Drug Resistant *Pseudomonas aeruginosa*." *Phytomedicine* 20(11):956-963. doi: 10.1016/j.phymed.2013.04.011.

Annapoorani Angusamy, Venugopal Umamageswaran, Radhakrishnan Parameswari, Shunmugiah Karutha Pandian, and Arumugam Veera Ravi. 2012b. "Computational Discovery of Putative Quorum Sensing Inhibitors against LasR and RhlR Receptor Proteins of *Pseudomonas aeruginosa.*" *Journal of Computer - Aided Molecular Design* 26:1067-1077. doi: 10.1007/s10822-012-9599-1.

Arakawa Yoshichika, Yasuyoshi Ike, Mitsuaki Nagasawa, Naohiro Shibata, Yohei Doi, Keigo Shibayama, Tetsuya Yagi, and Takeshi Kurata. 2000.

"Trends in Antimicrobial-Drug Resistance in Japan." *Emerging Infectious Diseases* 6:572-575. doi: 10.3201/eid0606.000604.

Austin Brain, and X-H. Zhang. 2006. "*Vibrio harveyi*: A Significant Pathogen of Marine Vertebrates and Invertebrates." *Letters in Applied Microbiology* 43:119-124. doi.org/10.1111/j.1472-765X.2006.01989.x.

Autret Nicolas, Catherine Raynaud, Iharilalao Dubail, Patrick Berche, and Alain Charbit. 2003. "Identification of the *agr* Locus of *Listeria monocytogenes*: Role in Bacterial Virulence." *Infection and Immunity* 71:4463-4471. doi:10.1128/IAI.71.8.4463-4471.2003.

Baldi Ashish. 2010. "Computational Approaches for Drug Design and Discovery: An Overview." *Systematic Reviews in Pharmacy*: 99-105. doi: 10.4103/0975-8453.59519.

Blundell Tom, Guy Dodson, Dorothy Hodgkin, and Dan Mercola. 1972. "*Insulin:* The *Struct*ure in the Crystal and its Reflection in Chemistry and Biology." *Advances in Protein Chemistry* 26:279-402. doi: 10.1016/S0065-3233(08)60143-6.

Borchardt Scott A., Eric J Allain, James J. Michels, G.W. Stearns, R.F. Kelly, and William F Mccoy. 2001. "Reaction of Acylated Homoserine Lactone Bacterial Signaling Molecules with Oxidized Halogen Antimicrobials." *Applied and Environmental Microbiology* 67:3174-3179. doi: 10.1128/AEM.67.7.3174-3179.2001.

Bovbjerg Rasmussen Thomas, Thomas Bjarnsholt, Mette Elena Skindersoe, Morten Hentzer, Peter Kristoffersen, Manuela Kote, John Nielsen, Leo Eberl, and Michael Givskov. 2005. "Screening for Quorum-Sensing Inhibitors (QSI) by Use of a Novel Genetic System, the QSI Selector." *Journal of Bacteriology* 187:1799-1814. doi: 10.1128/JB.187.5.1799-1814.2005.

Bowen William H. 2002. "Do We Need to be Concerned about Dental Caries in the Coming Millennium?." *Critical Reviews in Oral Biology and Medicine* 13:126-131. doi: 10.1177/154411130201300203.

Cady Nathaniel C., Kurt A. McKean, Jason Behnke, Roman Kubec, Aaron P. Mosier, Stephen H. Kasper, David S. Burz, and Rabi A. Musah. 2012. "Inhibition of Biofilm Formation, Quorum Sensing and Infection in *Pseudomonas aeruginosa* by Natural Products-Inspired Organosulfur

Compounds." PLoS ONE 7: e38492. doi: 10.1371/journal.pone. 0038492.

Caetano L., M. Antunes, Rosana B.R. Ferreira, Michelle M.C. Buckner, and B. Brett Finlay. 2010. "Quorum Sensing in Bacterial Virulence." *Microbiology* 156:2271-2282. doi: 10.1099/mic.0.038794-0.

Carlier Aurelien, Stephane Uroz, B. Smadja, Rupert G Fray, Xavier Latour, Yves Dessaux, and Denis Faure. 2003. "The Ti Plasmid of *Agrobacterium tumefaciens* Harbors an *attM* Paralogous Gene, *aiiB*, also Encoding *N*-Acyl Homoserine Lactonase Activity." *Applied and Environmental Microbiology* 69:4989-4993. doi: 10.1128/AEM.69.8. 4989-4993.2003.

Chbib Christiane. 2019. "Impact of the Structure-Activity Relationship of AHL Analogues on Quorum Sensing in Gram-Negative Bacteria." *Bioorganic and Medicinal Chemistry*. 28(3): 115282. doi: 10.1016/j. bmc.2019.115282.

Choo Jeong Han, Yaya Rukayadi, and J-K. Hwang. 2006. "Inhibition of Bacterial Quorum Sensing by Vanilla Extract." *Letters in Applied Microbiology* 42:637-641. doi: 10.1111/j.1472-765X.2006.01928.x.

Cos Paul, Arnold J Vlietinck, Dirk vanden berghe, and Louis Maes. 2006. Anti-Infective Potential of Natural Products: How to Develop a Stronger *In Vitro* "Proof of Concept." *Journal of Ethnopharmacology* 106:290-302. doi: 10.1016/j.jep.2006.04.003.

Coulthurst Sarah J., Neil R. Williamson, Abigail K.P. Harris, David R. Spring, and George P.C. Salmond. 2006. "Metabolic and Regulatory Engineering of *Serratia marcescens*: Mimicking Phage-Mediated Horizontal Acquisition of Antibiotic Biosynthesis and Quorum Sensing Capacities." *Microbiology* 152:1899-1911. doi: 10.1099/mic.0.28803-0.

Dahanukar Sharadini A., Rhushikesh A Kulkarni, and Nirmala N. Rege. 2000. "Pharmacology of Medicinal Plants and Natural Products." *Indian Journal of Pharmacology* 32: S81-S118.

Defoirdt Tom, Nico Boon, Patrick Sorgeloos, Willy Verstraete, and Peter Bossier. 2008. "Quorum Sensing and Quorum Quenching in *Vibrio harveyi*: Lessons Learned from *In vivo* Work." *ISME Journal* 2:19-26. doi: 10.1038/ismej.2007.92.

Defoirdt Tom. 2017. "Quorum-Sensing Systems as Targets for Antivirulence Therapy." *Trends in Microbiology*. doi: 10.1016/j.tim.2017.10.005.

Deziel Eric, Francois Lepine, Sylvain Milot, Jianxin He, Michael N Mindrinos, Ronald G Tompkins, and Laurence G Rahme. 2004. "Analysis of *Pseudomonas aeruginosa* 4-hydroxy-2-alkylquinolines (HAQs) Reveals a Role for 4-Hydroxy-2- Heptylquinoline in Cell-to-Cell communication." *Proceedings of the National Academy of Sciences*, USA 101:1339-1344. doi: 10.1073/pnas.0307694100.

Diacovich Lautaro, and Jean-Pierre Gorvel. 2010. "Bacterial Manipulation of Innate Immunity to Promote Infection." *Nature reviews Microbiology*. 8(2):117-128. doi: 10.1038/nrmicro2295.

Dong Yi-Hu, Jin-Ling Xu, Xian-Zhen Li, and Lian-Hui Zhang. 2000. "AiiA, an Enzyme that Inactivates the Acylhomoserine Lactone Quorum-Sensing Signal and Attenuates the Virulence of *Erwinia carotovora*." *Proceedings of the National Academy of Sciences USA* 97:3526-3531. doi: 10.1073/pnas.97.7.3526.

Donlan Rodney M., and J. William Costerton. 2002. "Biofilms: Survival Mechanisms of Clinically Relevant Microorganisms." *Clinical Microbiology Reviews* 15:167-193. doi: 10.1128/CMR.15.2.167-193.2002.

Eberl Leo. 1999. "N-Acyl Homoserine Lactone-Mediated Gene Regulation in Gram-Negative Bacteria." *Systematic and Applied Microbiology* 22(4):493-506. doi: 10.1016/S0723-2020(99)80001-0.

Galloway Darrell R. 1991. "*Pseudomonas aeruginosa* Elastase and Elastolysis Revisited: Recent Developments."*Molecular Microbiology* 5:2315-2321. doi: 10.1111/j.1365-2958.1991.tb02076.x.

Galloway Warren R.J.D., James T Hodgkinson, Steven D Bowden, Martin Welch, and David R Spring. 2011. "Quorum Sensing in Gram-Negative Bacteria: Small-Molecule Modulation of AHL and AI-2 Quorum Sensing Pathways." *Chemical reviews* 111:28-67. doi: 10.1021/cr100109t.

Genead Mohamed A., Jason J. McAnany, and Gerald A. Fishman. 2006. "Topical Dorzolamide for the Treatment of Cystoid Macular Edema in

Patients with Retinitis Pigmentosa." *American journal of ophthalmology* 141: 850-858. doi: 10.1097/IAE.0b013e3182215ae9.

Geske Grant D., Jennifer C.O. Neill, and Helen E. Blackwell. 2008. "Expanding Dialogues: From Natural Autoinducers to Non-Natural Analogues that Modulate Quorum Sensing in Gram-Negative Bacteria." *Chemical Society Reviews* 37:1432-1447. doi: 10.1039/b703021p.

Givskov Michael, Rocky De Nys, Michael Manefield, Lone Gram, Ria Maximilien, Leo Eberl, Soren Molin, Peter D. Steinberg, and Staffan Kjelleberg. 1996. "Eukaryotic Interference with Homoserine Lactone-Mediated Prokaryotic Signalling." *Journal of Bacteriology* 178:6618-6622. doi: 10.1128/jb.178.22.6618-6622.

Giwercman Birgit, Elsebeth Tvenstrup Jensen, Niels Hoiby, Arsalan Kharazmi, and John William Costerton. 1991. "Induction of Beta-Lactamase Production in *Pseudomonas aeruginosa* Biofilm." *Antimicrobial Agents and Chemotherapy* 35:1008-1010. doi: 10.1128/AAC.35.5.1008.

Goodgame J.C., J.C. Jr. Pottage, Helmut Jablonowski, W David Hardy, A. Stein, M. Fischl, P. Morrow, Judith Feinberg, Cindy H Brothers, I. Vafidis, P. Nacci, Jane Yeo, Louise Pedneault. 2000. "Amprenavir in Combination with Lamivudine and Zidovudine versus Lamivudine and Zidovudine alone in HIV-1-Infected Antiretroviral-Naive Adults. Amprenavir PROAB3001 International Study Team." *Antiviral therapy* 5:215-225. PMID: 11075942.

Greenberg Peter E, Peter E Greenberg, Bryce V Plapp, and Matthew R Parsek. 2000. "Autoinducer Synthase Modulating Compounds and Uses thereof." *PCT International Application.*

Hancock Robert E.W. 1998. "Resistance Mechanisms in *Pseudomonas aeruginosa* and other Nonfermentative Gram-Negative Bacteria." *Archives of clinical infectious diseases* 27(Suppl. 1): S93-S99. doi: 10.1086/514909.

Hejazi A., and F.R. Falkiner. 1997. "*Serratia marcescens.*" *Journal of medical microbiology* 46:903-912.doi: 10.1099/00222615-46-11-903.

Hentzer Morten and Michael Givskov. 2003. "Pharmacological Inhibition of Quorum Sensing for the Treatment of Chronic Bacterial Infections."

Journal of Clinical Investigation 112:1300-1307. doi: 10.1172/JCI20074.

Hojo Keiichi, Satoshi Nagaoka, Toshio Ohshima, and Norikazu Maeda. 2009. "Bacterial Interactions in Dental Biofilm Development." *Journal of dental research* 88:982-990. doi: 10.1177/0022034509346811.

Houdt Rob Van, Michael Givskov, and Chris W Michiels. 2007. "Quorum Sensing in *Serratia*." *FEMS Microbiology Reviews* 31:407-424. doi: 10.1111/j.1574-6976.2007.00071.x.

Hugh W. Acton, and R.N. Chopra. 1933. "The Treatment of Chronic Intestinal Amoebiasis with the Alkaloids of *Holarrhena antidysenterica* (kurchi)." *Indian Medical Gazette* 68:11-27.

Inchagova Ksenia, Galimzhan Duskaev, and D.G. Deryabin. 2019. "Quorum Sensing Inhibition in *Chromobacterium violaceum* by Amikacin Combination with Activated Charcoal or Small Plant-Derived Molecules (Pyrogallol and Coumarin)." *Microbiology* 88(1): 63-71. doi: 10.1134/S0026261719010132.

Ishida Hiroko, Yoshihisa Ishida, Yuichi Kurosaka, Tsuyoshi Otani, Kenichi Sato, and Hiroyuki Kobayashi. 1998. "*In vitro* and *In vivo* Activities of Levofloxacin against Biofilm- Producing *Pseudomonas aeruginosa*." *Antimicrobial Agents and Chemotherapy* 42:1641-1645. doi: 10.1128/AAC.42.7.1641.

Itzstein Mark von, Wen-Yang Wu, Gaik B. Kok, Michael S. Pegg, Jeffrey C. Dyason, Betty Jin, Tho Van Phan, Mark L. Smythe, Hume F. White, Stuart W. Oliver, Peter M. Colman, Joseph N. Varghese, D. Michael Ryan, Jacqueline M. Woods, Richard C. Bethell, Vanessa J. Hotham, Janet M. Cameron, and Charles R. Penn. 1993. "Rational Design of Potent Sialidase-Based Inhibitors of Influenza Virus Replication." *Nature* 363:418-423. doi: 10.1038/363418a0.

Jahoor Aruna, Rashila Patel, Amanda Bryan, Catherine Do, Jay Krier, Chase Watters, Walter Wahli, Guigen Li, Simon C. Williams, and Kendra P. Rumbaugh. 2008. "Peroxisome Proliferator Activated Receptors Mediate Host Cell Proinflammatory Responses to *Pseudomonas aeruginosa* Autoinducer." *Journal of Bacteriology* 190(13): 4408-4415. doi: 10.1128/JB.01444-07.

Jianga Kai, Xinlin Yan, Jiahao Yu, Zijian Xiao, Hao Wu, Meihua Zhao, Yuandong Yue, Xiaoping Zhou, Junhai Xiao, and Feng Lin. 2020. "Design, Synthesis, and Biological Evaluation of 3-amino-2-oxazolidinone Derivatives as Potent Quorum-Sensing Inhibitors of *Pseudomonas aeruginosa* PAO1." *European journal of medicinal chemistry* 194: 112-252. doi: 10.1016/j.ejmech.2020.112252.

Joint Ian, Karen Tait, Maureen E Callow, James A Callow, Debra Milton, Paul Williams, and Miguel Camara. 2002. "Cell-to-Cell Communication across the Prokaryote–Eukaryote Boundary." *Science* 298:1207. doi: 10.1126/science.1077075.

Kandasamy Saveetha, Wajahatullah Khan, Franklin Evans, Alan T. Critchley, and Balakrishnan Prithiviraj. 2012. "A Product of *Ascophyllum nodosum* Enhances Immune Response of *Caenorhabditis elegans* against *Pseudomonas aeruginosa* Infection." *Marine Drugs* 10:84-105. doi: 10.3390/md10010084.

Karlsson Thommie, Maria V. Turkina, Olena Yakymenko, Karl-Eric Magnusson, and Elena Vikstrom. 2012. "The *Pseudomonas aeruginosa* N-Acylhomoserine Lactone Quorum Sensing Molecules Target IQGAP1 and Modulate Epithelial Cell Migration." *PLoS Pathogens* 8(10):e1002953. doi: 10.1371/journal.ppat.1002953.

Kempne Ellis S., and Frank E Hanson. 1968. "Aspects of Light Production by *Photobacterium fischeri*." *Journal of Bacteriology* 95:975-979. doi: 10.2307/1540167.

Kernacki Karen A., Jeffery A. Hobden, Linda D Hazlett, Rafael A. Fridman, and Richard S. Berk. 1995. "*In Vivo* Bacterial Protease Production during *Pseudomonas aeruginosa* Corneal Infections." *Investigative Ophthalmology and Visual Science* 36:1371-1378. PMID: 7775115.

Kessler Efrat, Mary Israel, Nahum Landshman, Aaron Chechick, and Shmaryahu Blumberg. 1982. "*In Vitro* Inhibition of *Pseudomonas aeruginosa* Elastase by Metal-Chelating Peptide Derivatives." *Infection and Immunity* 38:716-723.

Khan Mohd Sajjad ahmad, Maryam Zahin S. Hasan, Fohad Mabood Husain, and Iqbal Ahmad. 2009. "Inhibition of Quorum Sensing Regulated Bacterial Functions by Plant Essential Oils with Special Reference to

Clove Oil." *Letters in Applied Microbiology* 49:354-360. doi: 10.1111/j. 1472-765X.2009.02666.x.

Kim Jung Sun, Yeon Hee Kim, Young Wan Seo and Sunghoon Park. 2007. "Quorum Sensing Inhibitors from the Red Alga, *Ahnfeltiopsis flabelliformis.*" *Biotechnology and Bioprocess Engineering* 12:308-311. doi: 10.1007/BF02931109.

Klebe Gerhard. 2006. "Virtual Ligand Screening: Strategies, Perspectives and Limitations."*Drug Discovery Today* 11:580-594. doi: 10.1016/j.drudis.2006.05.012.

Kolb Peter, Rafaela S. Ferreira, John J. Irwin, and Brian K. Shoichet. 2009. "Docking and Chemoinformatic Screens for New Ligands and Targets." *Current opinion in Biotechnology* 20:429-436. doi: 10.1016/j.copbio.2009.08.003.

Kolenbrander Paul E., Roxanna N Andersen, David S Blehert, Paul G. Egland, Jamie S. Foster, and Robert J. Palmer Jr. 2002. "Communication among Oral Bacteria*.*" *Microbiology and Molecular Biology Reviews* 66:486-505. doi: 10.1128/mmbr.66.3.486-505.2002.

Koo H. 2008. "Strategies to Enhance the Biological Effects of Fluoride on Dental Biofilms." *Journal of advanced dental research* 20:17-21. doi: 10.1177/154407370802000105.

Krishnan Thiba, Wai-Fong Yin, and Kok-Gan Chan. 2012. "Inhibition of Quorum Sensing-Controlled Virulence Factor Production in *Pseudomonas aeruginosa* PAO1 by Ayurveda Spice Clove (*Syzygium Aromati*cum) Bud Extract." *Sensors* 12:4016-4030. doi: 10.3390/s120404016.

Kumar Purnima S., Ann L. Griffen, Melvin L. Moeschberger, and Eugene J. Leys. 2005. "Identification of Candidate Periodontal Pathogens and Beneficial Species by Quantitative 16S Clonal Analysis." *Journal of Clinical Microbiology* 43:3944- 3955. doi: 0.1128/JCM.43.8.3944-3955.2005.

Latifi Amel, Michael K. Winson, Maryline Foglino, Barrie W. Bycroft, Gordon S.A.B. Stewart, Andree Lazdunski, and Paul Williams. 1995. "Multiple Homologues of LuxR and LuxI Control Expression of Virulence Determinants and Secondary Metabolites through Quorum

Sensing in *Pseudomonas aeruginosa* PAO1." *Molecular Microbiology* 17:333-343. doi: 10.1111/j.1365-2958.1995.mmi_17020333.x.

Leadbetter Jared R., and E. Peter Greenberg. 2000. Metabolism of Acyl-Homoserine Lactone Quorum-Sensing Signals by *Variovorax paradoxus*. *Journal of Bacteriology*182:6921-6926. doi: 10.1073/pnas. 96.24.13904.

Lewis Kim. 2001. Riddle of Biofilm Resistance. *Antimicrobial Agents and Chemotherapy*45:999-1007. doi: 10.1128/AAC.45.4.999-1007.2001.

Li J., E.J. Helmerhorst, Cataldo Leone, R.F. Troxler, T. Yaskell, A.D. Haffajee, S.S. Socransky, and Frank G Oppenheim. 2004. "Identification of Early Microbial Colonizers in Human Dental Biofilm." *Journal of Applied Microbiology* 97:1311-1318. doi: 10.1111/ j.1365-2672.2004.02420.x.

Liu Zunying, Yurong Pan, Xiaoshuang Li, Jinxin Jie, and Mingyong Zeng. 2017. "Chemical Composition, Antimicrobial and Anti-Quorum Sensing Activities of Pummelo Peel Flavonoid Extract." *Industrial Crops and Products* 109: 862-868. doi: 10.1016/j.indcrop.2017.09.054.

Maeda Hayato, Masashi Hosokawa, Tokutake Sashima, and Kazuo Miyashita. 2007. "Dietary Combination of Fucoxanthin and Fish Oil Attenuates the Weight Gain of White Adipose Tissue and Decreases Blood Glucose in Obese/Diabetic KK-Ay Mice." *Journal of Agricultural and Food Chemistry* 55:7701-7706. doi: 10.1021/ jf071569n.

Manefield Michael, Rocky de Nys, Kumar Naresh, Read Roger, Michael Givskov, Steinberg Peter, and Staffan Kjelleberg. 1999. "Evidence that Halogenated Furanones from *Delisea pulchra* Inhibit Acylated Homoserine Lactone (AHL)-Mediated Gene Expression by Displacing the AHL Signal from its Receptor Protein." *Microbiology* 145:283-291. doi: 10.1099/13500872-145-2-283.

Manefield Michael, Thomas Bovbjerg Rasmussen, Morten Henzter, Jens Bo Andersen, Peter Steinberg, Staffan Kjelleberg, and Michael Givskov. 2002. "Halogenated furanones inhibit quorum sensing through accelerated LuxR turnover." *Microbiology* 148:1119-1127. doi: 10. 1099/00221287-148-4-1119.

McGregor Maurice, and N. Segel. 1955. "The Rauwolfia Alkaloids in the Treatment of Hypertension." *British Heart Journal* 17:391-396. doi: 10.1136/hrt.17.3.391.

Miesel Lynn, Jonathan Greene, and Todd A. Black. 2003. "Microbial Genetics: Genetic Strategies for Antibacterial Drug Discovery." *Nature Reviews Genetics* 4:442-456. doi: 10.1038/nrg1086.

Miller Melissa B., and Bonnie L. Bassler. 2001. "Quorum Sensing in Bacteria." *Annual review of microbiology* 55:165-199. doi: 10.1146/annurev.micro.55.1.165.

Mohan S.B., S.C. Dinda, B.V.V.R. Kumar, J.R. Panda. 2012. "Computational Approaches for Drug Design and Discovery Process." *Current Pharmaceutical Research.* 2:600-611.

Musthafa Khadar Syed, Arumugam Veera Ravi, Angusamy Annapoorani, Sybiya Vasantha Packiavathy, and Shunmugiah Thevar Karutha Pandian. 2010. "Evaluation of Anti-Quorum-Sensing Activity of Edible Plants and Fruits through Inhibition of the N-Acyl-Homoserine Lactone System in *Chromobacterium violaceum* and *Pseudomonas aeruginosa.*" *Chemotherapy* 56:333-339. doi: 10.1159/000320185.

Musthafa Khadar Syed, Bhagavathi Sundaram Sivamaruthi, Shunmugiah Karutha Pandian, and Arumugam Veera Ravi. 2012. "Quorum Sensing Inhibition in *Pseudomonas aeruginosa* PAO1 by Antagonistic Compound Phenyl Acetic Acid." *Current Microbiology* 65:475-480. doi: 10.1007/s00284-012-0181-9.

Nealson Kenneth H. 1977. "Autoinduction of Bacterial Luciferase. Occurrence, Mechanism and Significance." *Archives of Microbiology* 112:73-79. doi: 10.1007/BF00446657.

Nealson Kenneth H., Terry Platt, and J. Woodland Hastings. 1968. "Bacterial Luciferase as an Inducible Enzyme." *Bacteriol Proc GP2*.

Nealson Kenneth H., Terry Platt, and J. Woodland Hastings. 1970. "Cellular Control of the Synthesis and Activity of the Bacterial Luminescent System." *Journal of Bacteriology* 104:313-322. doi: 10.1128/JB.104.1.313-322.

Ni Nanting, Minyong Li, Junfeng Wang, and Binghe Wang. 2009. "Inhibitors and Antagonists of Bacterial Quorum Sensing." *Medicinal research reviews* 29:65-124. doi: 10.1002/med.20145.

Nikaido Hiroshi. 1994. "Prevention of Drug Access to Bacterial Targets: Permeability Barriers and Active Efflux." *Science* 264:382-388. doi: 10.1126/science.8153625.

Niu C., S. Afre, and Eric S. Gilbert. 2006. "Sub Inhibitory Concentrations of Cinnamaldehyde Interfere with Quorum Sensing." *Letters in Applied Microbiology* 43:489-494. doi: 10.1111/j.1472-765X.2006.02001.x.

Ohtani Kaori, Yonghui Yuan, Sufi Hassan, Ruoyu Wang, Yun Wang, and Tohru Shimizu.2009. "Virulence Gene Regulation by the *agr* System in *Clostridium perfringens.*" *Journal of Bacteriology* 191:3919-3927. doi: 10.1128/JB.01455-08.

Ooms Frederic. 2000. "Molecular Modeling and Computer Aided Drug Design. Examples of Their Applications in, Medicinal Chemistry." *Current Medical Chemistry* 7:141-158. doi: 10.2174/092986 7003375317.

Sybiya Vasantha Packiavathy, Issac Abraham, Palani Agilandeswari, Khadar Syed Musthafa, Shunmugiah Karutha Pandian, and Arumugam Veera Ravi. 2012a. "Antibiofilm and Quorum Sensing Inhibitory Potential of *Cuminum cyminum* and its Secondary Metabolite Methyl Eugenol against Gram-Negative Bacterial Pathogens." *Food Research International* 45:85-92. doi: 10.1016/j.foodres.2011.10.022.

Sybiya Vasantha Packiavathy, Issac Abraham, Selvam Priya, Shunmugiah Karutha Pandian, and Arumugam VeeraRavi. 2012b. "Inhibition of Biofilm Development of Uropathogens by Curcumin – An Anti-Quorum Sensing Agent from *Curcuma longa."* *Food Chemistry* 148: 453-460. doi: 10.1016/j.foodchem.2012.08.002.

Park Sun-Yang, Sang Jun Lee, Tae-Kwang Oh, Jong-Won Oh, Bon-Tag Koo, Do-Young Yum, and Jung-Kee Lee. 2003. "AhlD, an N-Acylhomoserine Lactonase in *Arthrobacter* sp., and Predicted Homologues in Other Bacteria." *Microbiology* 149:1541-1550.

Parkins Michael D., Howard Ceri, and Douglas G. Storey. 2000. "*Pseudomonas aeruginosa* GacA, a Factor in Multihost Virulence, is also Essential for Biofilm Formation." *Molecular Microbiology* 40:1215-1226. doi: 10.1046/j.1365-2958.2001.02469.x.

Parsek Matthew R., Dale L. Val, Brian L. Hanzelka, John E. Cronan Jr., and E.P. Greenberg. 1999. "Acyl Homoserine-Lactone Quorum-Sensing Signal Generation." *Proceedings of the National Academy of Sciences* USA 96:4360-4365. doi: 10.1073/pnas.96.8.4360.

Patwardhan Bhushan, Ashok D.B. Vaidya, and Mukund Chorghade. 2004. "Ayurveda and Natural Products Drug Discovery." *Current Science* 86:789-799. https://www.jstor.org/stable/24109136.

Pesci Everett C., Jared B.J. Milbank, James P. Pearson, S.L. McKnight, Andrew S. Kende, E. Peter Greenberg, and Barbara H Iglewski. 1999. "Quinolone Signaling in the Cell-to-Cell Communication System of *Pseudomonas aeruginosa."* *Proceedings of the National Academy of Sciences* USA 96:11229-11234. doi: 10.1073/pnas.96.20.11229.

Peterson Lance W., and David Artis. 2014. "Intestinal Epithelial Cells: Regulators of Barrier Function and Immune Homeostasis." *Nature Reviews Immunology* 14(3):141-153. doi: 10.1038/nri3608.

Rahim Rahim, Urs A. Ochsner, Clarita Olvera, Michael Graninger, Paul Messner, Joseph S. Lam, and Gloria Soberon-Chavez. 2001. "Cloning and Functional Characterization of the *Pseudomonas aeruginosa rhlC* Gene that Encodes Rhamnosyltransferase 2, an Enzyme Responsible for Di-rhamnolipid Biosynthesis." *Molecular Microbiology* 40:708-718. doi: 10.1046/j.1365-2958.2001.02420.x.

Rasmussen Thomas B., and Michael Givskov. 2006. Quorum Sensing Inhibitors: A Bargin of Effects. *Microbiology* 152:895-904. doi: 10.1099/mic.0.28601-0.

Reen Jerry, Jose Antonio Gutierrez Barranquero, Maria Lopez Parages, and Fergal O'Gara. 2018. "Coumarin: A Novel Player in Microbial Quorum Sensing and Biofilm Formation Inhibition." *Applied Microbiology and Biotechnology* 102:2063–2073. doi: 10.1007/s00253-018-8787-x.

Riccioni Graziano, Barbara Mancini, E. Di llio, Tonino Bucciarelli, and Nicolantonio D'Orazio. 2008. "Protective Effect of Lycopene in Cardiovascular Disease." *European Review for Medical and Pharmacological Sciences* 12:183-190.

Ryan Robert P., and J. Maxwell Dow. 2008. "Diffusible Signals and Interspecies Communication in Bacteria." *Microbiology* 154:1845-1858. doi: 10.1099/mic.0.2008/017871-0.

Sachindra Nakkarike M., Emiko Sato, Hayato Maeda, Masashi Hosokawa, Yoshimi Niwano, Masahiro Kohno, and Kazuo Miyashita. 2007. "Radical Scavenging and Singlet Oxygen Quenching Activity of Marine Carotenoid Fucoxanthin and its Metabolites." *Journal of Agricultural and Food Chemistry* 55:8516-8522. doi: 10.1007/s13197-010-0022-4.

Schaefer Amy L., Dale L. Val, Brian L Hanzelka, John Cronan, and E Peter Greenberg. 1996. "of Cell-to-Cell Signals in Quorum Sensing: Acyl Homoserine Lactone Synthase Activity of a Purified *Vibrio fischeri* LuxI Protein." *Proceedings of the National Academy of Sciences USA* 93:9505-9509. doi: 10.1073/pnas.93.18.9505.

Schuster Martin, and E. Peter Greenberg. 2006. "A Network of Networks: Quorum-Sensing Gene Regulation in *Pseudomonas aeruginosa*." *International Journal of Medical Microbiology* 296:73-81. doi: 10.1016/j.ijmm.2006.01.036.

Senturk Sezgi, Seyhan Ulusoy, Gulgun Bosgelmez-Tinaz, and Aysegul Yagci. 2012. "Quorum Sensing and Virulence of *Pseudomonas aeruginosa* during Urinary Tract Infections." *Journal of Infection in Developing Countries* 6:501-507. doi: 10.3855/jidc.2543.

Shigeta M., G. Tanaka, H. Komatsuzawa, M. Sugai, H. Suginaka, and T. Usui. 1997. "Permeation of Antimicrobial Agents through *Pseudomonas aeruginosa* Biofilms: A Simple Method." *Chemotherapy* 43:340-345. doi: 10.1159/000239587.

Shiratori Kenji, Kazuhiro Ohgami, Iliyana Ilieva, Xue-HaiJin, Yoshikazu Koyama, Kazuo Miyashita, KazuhikoYoshida, Satoru Kase, and Shigeaki Ohnoa.2005. "Effects of Fucoxanthin on Lipopolysaccaride-Induced Inflammation *In vitro* and *In Vivo*." *Experimental Eye Research* 81:422-428. doi: 10.1016/j.exer.2005.03.002.

Sifri Costi D. 2008. "Healthcare Epidemiology: Quorum Sensing: Bacteria Talk Sense." *Clinical Infectious Diseases* 47:1070-1076. doi: 10.1086/592072.

Sio Charles F., Linda G. Otten, Robbert H. Cool, Stephen P. Diggle, Peter G. Braun, Rein Bos, Mavis Daykin, Miguel Camara, Paul Williams, and Wim J. Quax. 2006. "Quorum Quenching by an *N*-Acyl-Homoserine Lactone Acylase from *Pseudomonas aeruginosa* PAO1." *Infection Immunity* 74:1673-1682. doi: 10.1128/IAI.74.3.1673-1682.2006.

Skindersoe Mette Elena, Piers Ettinger-Epstein, Thomas Bovbjerg Rasmussen, Thomas Bjarnsholt, Rocky de Nys, and Michael Givskov. 2008. "Quorum Sensing Antagonism from Marine Organisms." *Marine Biotechnology* 10:56-63. doi: 10.1007/s10126-007-9036-y.

Smith Roger S., and Barbara H. Iglewski. 2003. "*Pseudomonas aeruginosa* Quorum Sensing as a Potential Antimicrobial Target." *Journal of Clinical Investigation* 112:1460-1465. doi: 10.1172/JCI20364.

Snodderly D. Max. 1995. "Evidence for Protection against Age-Related Macular Degeneration by Carotenoids and Antioxidant Vitamins." *American Journal of Clinical Nutrition* 62:1448S-1461S. doi: 10.1093/ajcn/62.6.1448S.

Song Chun Meng, Shen Jean Lim, and Joo Chuan Tong. 2009. "Recent Advances in Computer-Aided Drug Design." *Briefings in Bioinformatics* 10:579-591. doi: 10.1093/bib/bbp023.

Spirt Silke D, Kaya Lutter, Wilhelm Stahl. 2010. "Carotenoids in Photooxidative Stress." *Current Nutrition and Food Science* 6:36-43. doi: 10.2174/157340110790909572.

Stewart Philip S. 1996. "Theoretical Aspects of Antibiotic Diffusion into Microbial Biofilms." *Antimicrobial Agents and Chemotherapy* 40:2517-2522. doi: 10.1128/AAC.40.11.2517.

Stock Ingo, Sonja Burak, Kimberley Jane Sherwood, Thomas Gruger, and Bernd Wiedemann. 2003. "Natural Antimicrobial Susceptibilities of Strains of 'Unusual' *Serratia* Species: *S. ficaria, S. fonticola, S. odorifera, S. plymuthica* and *S. rubidaea.*" *Journal of Antimicrobial Chemotherapy* 51:865-885. doi: 10.1093/jac/dkg156.

Suh Sang-Jin, Laura Silo-Suh, Donald E. Woods, Daniel J. Hassett, Susan E.H. West, and Dennis E. Ohman. 1999. "Effect of *rpoS* Mutation on the Stress Response and Expression of Virulence Factors in

Pseudomonas aeruginosa." *Journal of Bacteriology* 181:3890-3897. doi: 10.1128/JB.181.13.3890-3897.1999

Sybiya Vasantha Packiavathy, Issac Abraham, Agilandeswari Palani, Babu Rajendran Ramaswamy, Karutha Pandian Shunmugiah, and Veera Ravi Arumugam. 2011. "Antiquorum Sensing and Antibiofilm Potential of *Capparis spinose.*" *Archives of Medical Research* 42:658-668. doi: 10.1016/j.arcmed.2011.12.002.

Terstappen Georg C., and Angelo Reggiani. 2001. "*In silico* Research in Drug Discovery." *Trends in pharmacological sciences* 22:23-26. doi: 10.1016/S0165-6147(00)01584-4.

Tinaz Gulgun Bosgelmez. 2003. "Quorum Sensing in Gram-Negative Bacteria." *Turkish Journal of Biology* 27:85-93.

Turkina Maria V., and Elena Vikstrom. 2019. Bacteria-Host Crosstalk: Sensing of the Quorum in the Context of *Pseudomonas aeruginosa* Infections. *Journal of Innate Immunity* 11:263-279. doi: 10.1159/000494069.

Uroz Stephane, Cathy D'Angelo-Picard, Aurélien Carlier, Miena Elasri, Carine Sicot, Annik Petit, Phil Oger, Denis Faure, and Yves Dessaux. 2003. "Novel Bacteria Degrading N-Acyl Homoserine Lactones and Their Use as Quenchers of Quorum-Sensing-Regulated Functions of Plant-Pathogenic Bacteria." *Microbiology* 149:1981-1989. doi: 10.1099/mic.0.26375-0.

Uroz Stephane, Siri Ram Chhabra, Miguel Camara, Paul Williams, Phil Oger, and Yves Dessaux. 2005. "N-Acylhomoserine Lactone Quorum-Sensing Molecules are Modified and Degraded by *Rhodococcus erythropolis* W2 by Both Amidolytic and Novel Oxidoreductase Activities." *Microbiology* 151:3313-3322. doi: 10.1099/mic.0.27961-0.

Van Delden Christian, and Barbara H. Iglewski. 1998. "Cell-to-Cell Signaling and *Pseudomonas aeruginosa* Infections." *Emerging Infectious Disease* 4:551-560. doi: 10.3201/eid0404.980405.

Vikram Amit, Guddadarangavvanahally K. Jayaprakasha, Palmy Jesudhasan, Suresh D Pillai, and Bhimanagouda S Patil. 2010. "Suppression of Bacterial Cell–Cell Signalling, Biofilm Formation and

Type III Secretion System by Citrus Flavonoids." *Journal of Applied Microbiology* 109:515-527. doi: 10.1111/j.1365-2672.2010.04677.x.

Vikstrom Elena, Lan Bui, Peter Konradsson, and Karl-Eric Magnusson. 2010. Role of Calcium Signalling and Phosphorylations in Disruption of the Epithelial Junctions by *Pseudomonas aeruginosa* Quorum Sensing Molecule. *European Journal of Cell Biology* 89(8): 584-597. doi: 10.1016/j.ejcb.2010.03.002.

Vivek Prashanth Kumar, Neelam S. Chauhan, Harish Padh, and M. Rajani. 2006. "Search for Antibacterial and Antifungal Agents from Selected Indian Medicinal Plants." *Journal of Ethnopharmacology*107:182-188. doi: 10.1016/j.jep.2006.03.013.

Wang Rui, Melissa Starkey, Ronen Hazan, and Laurence G. Rahme. 2012. "Honey's Ability to Counter Bacterial Infections Arises from both Bactericidal Compounds and QS Inhibition." *Frontiers in Microbiology* 3:144. doi: 10.3389/fmicb.2012.00144.

Waters Christopher M. and Bonnie L. Bassler. 2005. "Quorum Sensing: Cell-to-Cell Communication in Bacteria." *Annual Review of Cell and Developmental Biology.* 21:319-346. doi: 10.1146/annurev.cellbio.21.012704.131001.

Wei Jun-Rong, and Hsin-Chih Lai. 2006. "N-Acylhomoserine Lactone-Dependent Cell-to-Cell Communication and Social Behavior in the Genus *Serratia."* *International Journal of Medical Microbiology* 296:117-124. doi: 10.1016/j.ijmm.2006.01.033.

Whitehead Neil A., Anne M.L. Barnard, Holly Slater, Natalie J.L. Simpson, and George P.C. Salmond. 2001. "Quorum-Sensing in Gram-Negative Bacteria." *FEMS Microbiology Reviews* 25:365-404. doi: 10.1111/j.1574-6976.2001.tb00583.x.

Wick M.J., A.N. Hamood, and B.H. Iglewski. *1990.* "Analysis of the Structure-Function Relationship of *Pseudomonas aeruginosa* Exotoxin A." *MolecularMicrobiology* 4:527-535. doi: 10.1111/j.1365-2958.1990.tb00620.x

Yang Liang, Morten Theil Rybtke, Tim Holm Jakobsen, Morten Hentzer, Thomas Bjarnsholt, Michael Givskov, and TimTolker-Nielsen.2009. "Computer-Aided Identification of Recognized Drugs as *Pseudomonas aeruginosa* Quorum-Sensing Inhibitors."*Antimicrobial Agents and Chemotherapy* 53: 2432-2443. doi: 10.1128/AAC.01283-08.

Yi-Han Lin, Jin-Ling Xu, Jiangyong Hu, Lian-Hui Wang, Say Leong Ong, Jared Renton Leadbetter, and Lian-Hui Zhang. 2003. "Acyl-Homoserine Lactone Acylase from *Ralstonia* Strain XJ12B Represents a Novel and Potent Class of Quorum-Quenching Enzymes." *Molecular Microbiology* 47:849-860. doi: 10.1046/j.1365-2958.2003.03351.x.

Yuan Yao, Eric A Berg, Catherine E Costello, Robert F Troxler, and Frank G Oppenheim. 2003. "Identification of Protein Components in Human Acquired Enamel Pellicle and Whole Saliva Using Novel Proteomics Approaches." *Journal of Bioloical Chemistry* 278:5300-5308. doi: 10.1074/jbc.M206333200.

Zeng Zhirui, Li Qian, Lixiang Cao, Hongming Tan, Yali Huang, Xiaoli Xue, Yong Shen, and Shining Zhou. 2008. "Virtual Screening for Novel Quorum Sensing Inhibitors to Eradicate Biofilm Formation of *Pseudomonas aeruginosa*." *Applied Microbiology and Biotechnology* 79:119-126. doi: 10.1007/s00253-008-1406-5.

ABOUT THE EDITORS

Jayapradha Ramakrishnan, PhD
Senior Assistant Professor
Centre for Research in Infectious Diseases (CRID)
School of Chemical and Biotechnology
SASTRA Deemed to be University
Tanjavur, India

Jayapradha Ramakrishnan is presently working as Senior Assistant Professor at SASTRA Deemed to be University. She has obtained her Ph.D in the field of Microbiology and continuing research with a major focus in Antimicrobial Chemotherapy. Her research interest includes the strategies to tackle *Cryptococcus neoformans* and multi drug resistant *Klebsiella* spp, which made her obtain sponsored projects from the Government of India and has authored several scientific papers and book chapters.

Thiagarajan Raman, PhD
Assistant Professor and Head
Department of Advanced Zoology and Biotechnology
Ramakrishna Mission Vivekananda College
Chennai, India

Thiagarajan Raman is an Assistant Professor of Zoology and has over 12 years of research experience in the field of natural products, immunotoxicology, oxidative stress biology and inflammation and has authored several scientific papers and book chapters on these lines (https://scholar.google.com/citations?hl=en&user=KxeTUjQAAAAJ). This book is his first as an editor and explores the potential of natural compounds as antimicrobial agents.

INDEX

A

Abraham, 65, 69, 94, 95, 102, 161, 176, 180
acid, 6, 8, 10, 11, 27, 63, 92, 96, 98, 147, 158, 160, 164
adaptive immune response, 4, 11, 13, 19
adhesion, 56, 127, 153
ADP, 107, 133, 136
algae, 15, 16, 20
alkaloids, 24, 53, 115, 116, 159
alveolar macrophage, 57, 108, 111, 134, 142
amino, 10, 11, 12, 18, 113, 130, 131, 157, 162, 172
amino acid, 11, 12, 18, 113, 130
aminoglycosides, 53, 60, 61, 63, 65, 113, 130, 131
antibiotic(s), viii, 2, 22, 27, 29, 50, 53, 61, 63, 65, 92, 93, 94, 95, 99, 100, 101, 102, 113, 126, 127, 131, 132, 135, 139, 140, 141, 144, 145, 151
antibiotic resistance, v, viii, vii, 4, 22, 24, 27, 29, 30, 40, 50, 53, 55, 56, 58, 62, 66, 67, 69, 76, 91, 92, 93, 94, 99, 100, 101, 102, 103, 104, 106, 112, 113, 114, 119, 121, 126, 127, 129, 130, 131, 132, 135, 138, 139, 140, 141, 151, 153, 165
antibody, 4, 6, 15, 108
antigen, 6, 109, 128
antioxidant, 6, 7, 8, 9, 14, 16, 54, 66, 114, 159
apoptosis, 111, 134, 135
autoimmune disease, 3, 4, 6
avoidance, 99, 108, 115

B

bacteremia, 62, 91, 106
bacteria, 8, 12, 13, 18, 19, 22, 23, 26, 27, 28, 29, 31, 40, 42, 45, 50, 54, 56, 57, 58, 66, 70, 72, 77, 78, 81, 82, 86, 93, 94, 98, 99, 102, 103, 104, 105, 106, 107, 108, 109, 110, 112, 114, 115, 126, 129, 130, 131, 132, 133, 134, 135, 144, 146, 148, 149, 151, 153, 155, 158, 160, 168, 169, 170, 173, 175, 176, 178, 180, 181
bacterial infection, 27, 58, 66, 92, 108, 110, 156, 160
bacterial pathogens, 23, 29, 92, 139, 140, 144, 151, 155, 156, 160, 161, 165

bacteriophage, 50, 56, 58, 69, 86
bacterium, 50, 53, 106, 111, 129, 133, 144
bioactive compound, 2, 6, 7, 10
biomaterials, 2, 10
biomolecules, vi, 24, 105, 106, 107, 108, 109, 110, 113, 114, 115, 116, 117, 121, 122
blood, 6, 8, 17, 25, 155
bloodstream, 50, 63, 89, 103, 146
bone, 6, 15, 20
bone marrow, 6, 15, 20
breakdown, 56, 61, 161
burn, 58, 59, 146

C

Ca^{2+}, 11, 14, 16, 18
cancer, 2, 7, 9, 10, 17, 27, 51, 92, 159
capsular-polysaccharides, 50
carbohydrates, 16, 20, 107
cardiotonic, 7, 9, 17
cell death, 54, 125, 127, 155
chemical(s), 94, 95, 106, 108, 157, 163
chromosome, 92, 95, 99
chymotrypsin, 14, 16, 20
classes, vii, 96, 102
classification, 95, 96, 97
clinical trials, 112, 113, 116
colonization, 28, 51, 52, 152, 155
combination therapy, 61, 88, 89
community/communities, 49, 51, 53, 66, 131, 152, 153
complement, 5, 8, 11, 107, 109
compounds, viii, 2, 4, 6, 7, 9, 10, 22, 23, 50, 54, 68, 92, 93, 114, 116, 151, 156, 157, 158, 159, 160, 163, 164
conjugation, 93, 94, 144, 145
cost, 7, 60, 100, 132
curcumin, 5, 6, 113, 161
cure, 2, 3, 5, 6, 7, 60, 116, 163
cysteine, 12, 13, 14, 161
cystic fibrosis, 114, 135, 140, 141
cytokines, 4, 6, 7, 11, 12, 13, 15, 18, 51
cytotoxicity, 110, 136, 137

D

damages, 2, 4, 6
defence, 24, 107, 109, 115, 135, 155
degradation, 56, 94, 110, 111, 126, 128, 133, 134, 151, 157
derivatives, 8, 68, 157, 161
destruction, 94, 95, 128, 151, 157
detection, 92, 100, 101, 134
diseases, vii, 1, 2, 3, 4, 7, 9, 17, 22, 33, 39, 41, 44, 50, 51, 70, 71, 74, 75, 76, 79, 82, 107, 116, 117, 121, 122, 126, 146, 151, 152, 155, 159, 163, 167, 170, 178, 183
distribution, 65, 153, 163
DNA, 67, 72, 94, 122, 130, 138, 140, 154
dosage, 60, 87, 88, 153
drug resistance, 106, 126, 129, 130, 131, 132
drugs, 2, 3, 4, 6, 7, 25, 50, 52, 60, 61, 63, 64, 65, 66, 67, 106, 108, 114, 131, 159, 162, 164, 165

E

E. coli, 13, 25, 56, 59, 100, 160
efflux pump, 92, 94, 98, 99, 103, 109, 130, 139, 140
elongation, 101, 107, 155
Enterobacteriaceae, 49, 50, 61, 63, 71, 73, 78, 79, 80, 81, 88, 97, 103
environment(s), 94, 107, 135, 145, 153, 160
enzyme(s), 10, 28, 63, 64, 72, 92, 93, 94, 95, 96, 97, 98, 102, 108, 110, 112, 114, 128, 130, 131, 140, 150, 151, 155, 156, 157, 158
epithelial cells, 107, 113, 126, 127, 133, 135, 137, 139

epithelium, 109, 126, 155
EPS, 87, 151, 153
extracts, 8, 25, 27, 29, 50, 53, 55, 66, 161

F

flavonoids, 6, 9, 24, 28, 52, 53, 55
fluoroquinolones, 53, 61, 130, 136, 140, 151
food, 7, 53, 101, 116
formation, 14, 17, 55, 111, 113, 126, 130, 131, 133, 134, 144, 145, 146, 151, 153, 155, 157, 161
France, 70, 81, 117
fruits, 54, 161, 164
fungus/fungi, 8, 12, 13, 19, 23, 27, 54

G

gene expression, 133, 144, 149, 156, 160
genes, 92, 93, 94, 101, 127, 128, 129, 131, 133, 135, 136, 140, 141, 144, 146, 150, 151, 154, 156, 157, 158, 162, 165
genome, 135, 137, 141, 163
growth, 8, 10, 22, 26, 54, 55, 107, 110, 133, 151, 156

H

health, 7, 50, 53, 56, 60, 102, 114
HIV, 26, 42, 44, 160, 164, 170
host, 2, 4, 5, 6, 7, 10, 11, 12, 13, 14, 18, 19, 106, 107, 109, 110, 111, 115, 125, 126, 127, 133, 134, 135, 140, 144, 153, 154, 155, 160
human, 1, 4, 26, 29, 49, 91, 109, 110, 112, 113, 116, 126, 133, 139, 144, 154, 155, 159
hydrolysis, 8, 14, 15, 16, 20, 22
hydrolyzing enzymes, 92
hydroxyl, 8, 12, 131

hypertension, 9, 14, 159

I

IFN, 11, 15, 18, 110
immune function, 6, 8, 106, 114
immune response, 2, 3, 4, 5, 6, 7, 10, 12, 13, 15, 18, 20, 109, 110, 115, 134
immune system, 1, 2, 5, 6, 7, 12, 57, 108, 109, 114, 115, 153
immunity, 2, 3, 5, 8, 10, 11, 12, 13, 14, 18, 19, 32, 33, 34, 36, 43, 71, 73, 106, 108, 114, 116, 117, 118, 120, 121, 136, 139, 142, 167, 169, 172, 179, 180
immunocompromised, 2, 91, 106, 126
immunomodulation, vi, 2, 3, 4, 6, 45, 105, 106, 114, 115, 116
immunomodulator, 4, 17, 18
immunomodulatory, 2, 4, 6, 7, 8, 9, 10, 11, 12, 13, 14, 15, 16, 21, 22, 57, 108, 111, 114, 115
immunostimulatory, 9, 10, 16, 17
in vitro, 6, 8, 12, 19, 108, 110, 113, 115, 116, 137
in vivo, 6, 12, 19, 93, 108, 113, 134
India, 1, 8, 23, 24, 41, 96, 105, 125, 126, 138, 143, 183
individuals, 2, 27, 59, 61, 68, 88, 91, 106, 145
induction, 9, 17, 125, 127
industry/industries, 54, 159, 162
infection, 4, 5, 11, 13, 14, 18, 19, 22, 50, 52, 56, 58, 61, 64, 65, 66, 69, 89, 92, 102, 103, 108, 115, 116, 117, 127, 131, 137, 138, 147, 153, 155, 156, 161, 164
inflammation, 5, 57, 135, 139, 183
inhibition, 16, 53, 55, 100, 113, 144, 155
inhibitor, 28, 60, 62, 67, 110, 113, 156, 158
injury, 7, 57, 109, 136
innate immune response, 3, 9, 11, 112, 154

innate immunity, 10, 11, 12, 13, 14, 18, 106, 108

J

Japan, 8, 96, 97, 167

K

kill, 3, 11, 132
Klebsiella, iv, v, 5, 22, 23, 26, 29, 34, 35, 40, 49, 50, 52, 54, 58, 59, 62, 64, 65, 70, 71, 72, 73, 74, 75, 76, 77, 78, 79, 80, 81, 86, 87, 89, 91, 92, 93, 97, 102, 103, 145, 166, 183
Klebsiella pneumoniae, v, 5, 22, 49, 50, 89, 91, 93, 97, 102, 103, 166

L

lead, 106, 115, 116, 126, 130, 134, 154, 155, 159, 163, 164
lipids, 10, 54, 107
Listeria monocytogenes, 123, 145, 167
lymphocytes, 3, 8, 11, 12, 15, 21, 22, 115

M

macrophages, 7, 9, 10, 11, 15, 16, 18, 20, 34, 57, 65, 71, 106, 108, 109, 110, 111, 112, 114, 115, 117, 118, 120, 121, 122, 123, 134, 142
management, 52, 68, 92, 138, 163
matrix, 55, 131, 137, 151
medical, 22, 25, 92, 170
medicine, viii, 6, 8, 23, 24, 26, 27, 53, 54, 148, 152, 159
Mediterranean, 12, 13, 19, 35, 37
meningitis, 50, 66, 106
metabolism, 12, 109, 148, 160, 163
metabolites, 53, 145, 148, 153
mice, 5, 15, 16, 22, 58, 59, 86, 108, 111
microorganisms, 2, 4, 7, 8, 10, 14, 69, 70, 71, 139, 159
models, 4, 69, 113, 115, 116
molecules, viii, 3, 12, 29, 30, 54, 115, 116, 127, 132, 145, 150, 151, 154, 156, 157, 162, 163
morbidity, 22, 93, 106, 138
mortality, 51, 60, 61, 93, 102, 106, 114, 155
mutation(s), 2, 93, 94, 113, 130, 140, 151

N

National Academy of Sciences, 122, 169, 177, 178
natural compound, viii, 6, 144, 160, 164, 165, 184
natural compounds, viii, 6, 144, 160, 164, 165, 184
neutrophils, 6, 11, 134
nosocomial pneumonia, 50, 108, 146

O

ocular infections, 125, 126, 128, 130, 134
organ, 3, 4, 51, 92, 108
organism, 55, 93, 94, 100, 125
oxygen, 10, 55, 67
oyster(s), 13, 16, 19, 20, 21, 22

P

pathogenesis, 3, 108, 127, 147, 153
pathogens, vii, 5, 23, 25, 27, 29, 65, 68, 92, 107, 112, 129, 132, 133, 134, 135, 144, 148, 152, 162, 165
pathway(s), 8, 11, 16, 107, 108, 109, 110, 114, 115, 134, 135, 157, 158
PCR, 92, 100, 101

Index

penicillin, 98, 102, 132, 136, 151
pepsin, 14, 16, 20
peptide(s), 2, 4, 10, 12, 13, 14, 16, 20, 43, 107, 145, 153, 158
peripheral blood, 15, 20, 110
permeability, 94, 98, 151
phage, 52, 55, 72, 86, 87, 106, 113, 114
phagocytosis, 5, 6, 7, 11, 14, 18, 19, 64, 106, 108, 109, 110, 112, 114, 116, 117, 118, 120, 121, 122, 134, 137
pharmaceutical(s), 2, 24, 151, 159, 162
phenotype(s), 60, 112, 140
plants, 2, 8, 10, 11, 13, 17, 23, 24, 26, 28, 29, 50, 53, 54, 59, 85, 114, 159, 161, 164
plasmid, 92, 94, 95, 97, 142
pneumonia, 5, 51, 55, 56, 59, 60, 62, 64, 65, 66, 68, 87, 88, 91, 106, 114, 140
polysaccharide, 5, 55, 109, 131
polysaccharides, 2, 5, 9, 10, 50, 59, 62, 114, 115, 153
Polysaccharides, 9, 18, 40
population, 7, 53, 93, 131, 132, 133, 138, 151, 152
prevention, 3, 59, 101, 109, 116, 158
probability, viii, 29, 30
pro-inflammatory, 11, 13, 15, 18, 51, 111
proliferation, 8, 9, 12, 15, 16, 17, 18, 20, 21, 22, 128, 132
protection, 17, 58, 66
proteins, 10, 18, 107, 114, 127, 141, 145, 150, 154, 156, 157, 158, 164
Pseudomonas, vi, 23, 56, 69, 74, 75, 97, 105, 106, 110, 113, 116, 117, 118, 119, 120, 121, 122, 123, 125, 126, 127, 133, 135, 136, 137, 138, 139, 140, 141, 142, 145, 146, 147, 165, 166, 167, 169, 170, 171, 172, 173, 174, 175, 176, 177, 178, 179, 180, 181, 182
Pseudomonas aeruginosa, vi, 56, 105, 106, 110, 116, 117, 118, 119, 120, 121, 122, 123, 125, 126, 127, 133, 135, 136, 137, 138, 139, 140, 141, 142, 146, 165, 166, 167, 169, 170, 171, 172, 173, 174, 175, 176, 177, 178, 179, 180, 181, 182
pumps, 94, 99, 103, 109, 130, 139

Q

quercetin, 55, 114, 161
quorum sensing, 113, 144, 156, 174
quorum sensing inhibition, 144

R

receptor(s), 9, 11, 57, 106, 108, 109, 112, 115, 145, 147, 150, 154, 156, 158, 164
recognition, 11, 18, 57, 108, 115, 126, 133, 139, 150, 158
relevance, viii, 2, 130
replication, 130, 133, 135
researchers, viii, 2, 10, 22, 158, 160
resistance, vii, 3, 22, 27, 29, 50, 51, 52, 53, 55, 56, 58, 59, 60, 61, 62, 66, 67, 88, 89, 91, 92, 93, 94, 95, 96, 99, 100, 101, 102, 103, 104, 106, 109, 111, 112, 113, 114, 126, 127, 129, 130, 131, 134, 136, 137, 138, 139, 140, 141, 151, 152, 153, 155
response, 2, 3, 4, 5, 6, 7, 11, 13, 14, 16, 18, 19, 20, 51, 57, 87, 106, 108, 111, 112, 114, 125, 127, 135, 155
rhamnolipid, 147, 155, 177
root(s), 17, 54, 83, 84

S

secretion, 127, 136, 137, 139, 155
sensing, 57, 113, 144, 156, 174
sensitivity, 62, 67, 100, 132
showing, 24, 59, 85
side effects, 2, 7, 27, 53, 116, 159, 163
skin, 25, 51, 91
solution, 29, 65, 113

species, 4, 8, 17, 23, 24, 29, 51, 53, 54, 55, 57, 67, 91, 94, 145, 146, 152, 158
state, 6, 67, 132, 133, 144
stimulation, 4, 6, 57, 112, 115
structure, 56, 68, 102, 109, 156, 158, 163, 164
substrate, 99, 130, 153
Sun, 32, 35, 46, 76, 134, 173, 176
suppression, 3, 6, 15
survial statergies, 126
survival, 59, 62, 65, 68, 86, 107, 112, 126, 133, 134, 135, 162
susceptibility, 2, 26, 54, 63, 65, 93, 100
synthesis, 30, 58, 67, 128, 132, 147, 150, 153, 156, 157

T

T cell(s), 15, 16, 108
T lymphocytes, 3, 9, 10, 11, 15, 16, 18, 20
target, 6, 12, 94, 95, 99, 103, 107, 126, 129, 130, 132, 140, 145, 150, 154, 156, 162, 163
testing, 27, 59, 86
therapeutic approaches, 52, 114, 152
therapy, 4, 6, 8, 57, 60, 61, 67, 68, 72, 112, 113, 126, 170
tissue, 7, 11, 18, 59, 125, 128, 135, 155
TNF, 6, 9, 11, 13, 15, 16, 17, 18
toxicity, 6, 7, 10, 30, 61, 116
toxin, 127, 133, 155
transcription, 130, 132, 145, 154
transmission, 50, 52, 97
transplantation, 3, 51, 92
transport, 10, 55, 107, 127
treatment, 3, 4, 26, 50, 56, 58, 59, 60, 61, 62, 65, 66, 68, 69, 87, 88, 92, 106, 107, 109, 110, 112, 113, 126, 129, 132, 133, 151, 159, 164
trial, 2, 62, 66, 114, 136
trypsin, 14, 15, 16, 20

U

urinary tract, 27, 28, 50, 87, 88, 140, 146, 148
urinary tract infection, 27, 28, 50, 87, 140, 146, 148
USA, 76, 142, 169, 177, 178

W

worldwide, vii, 22, 30, 60, 62, 68, 92, 96, 106, 152
wound infection, 50, 58, 59, 146, 148